SPYING ON WHALES

SPYING

ON

WHALES

THE PAST, PRESENT, AND
FUTURE OF EARTH'S
MOST AWESOME CREATURES

NICK PYENSON

VIKING

VIKING

An imprint of Penguin Random House LLC
375 Hudson Street
New York, New York 10014
penguin.com

Illustrations by Alex Boersma. Illustrations © 2017 Alex Boersma

ISBN 9780735224568 (hardcover)
ISBN 9780735224575 (ebook)

Printed in the United States of America
1   3   5   7   9   10   8   6   4   2

Set in Sabon LT Std
Designed by Cassandra Garruzzo

Every author writes with a very specific reader in mind.
I wrote this book for you.
And for my family.

What have we been doing all these centuries but trying to call God back to the mountain, or, failing that, raise a peep out of anything that isn't us? What is the difference between a cathedral and a physics lab? Are not they both saying: Hello? We spy on whales and on interstellar radio objects; we starve ourselves and pray till we're blue.

—Annie Dillard, *Teaching a Stone to Talk*

For the animal shall not be measured by man. In a world older and more complete than ours, they move finished and complete, gifted with the extension of the senses we have lost or never attained, living by voices we shall never hear. They are not brethren, they are not underlings: they are other nations, caught with ourselves in the net of life and time, fellow prisoners of the splendour and travail of the earth.

—Henry Beston, *The Outermost House*

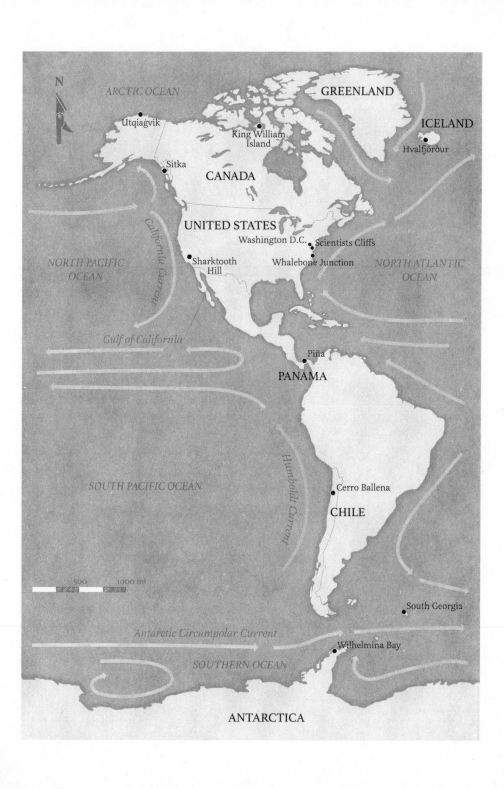

CONTENTS

SPYING ON WHALES

PROLOGUE

At this very moment, two spacecraft move at over thirty-four thousand miles per hour, about ten billon miles away from us, each carrying a gold-plated copper record. The spacecraft, *Voyager 1* and *Voyager 2*, are meant as messengers: they carry information about our address in the solar system, the building blocks of our scientific knowledge, and a small sampling of images, music, and greetings from around the world. They also carry whalesong.

The long squeaks and moans on the record belong to humpback whales. In the 1970s, when the Voyager mission launched, our view of whales was rapidly changing, from game animals to cultural icons and symbols of a nascent environmental movement. Scientists had recently discovered that male humpbacks produce complex songs, composed of phrases collected under broader themes, nested like Russian dolls that repeat in a loop. Humpback whalesong has evolved even since we started listening as each new singer improvises on the loop, creating new structures and hierarchies that constantly change over years, and across ocean basins.

Whalesong, however, remains a riddle to anyone who isn't a humpback whale. We can capture its variation, details, and complexity, but we don't know what any of it truly means. We lack the requisite context to decipher and understand it—or, really, any part of cetacean culture. Even so, we send whalesong into interstellar space because the creatures that sing these songs are superlative beings that fill us with awe, terror, and affection. We have hunted them for thousands of years and scratched them into our mythologies and iconography. Their bones frame the archways of medieval castles. They're so compelling that we imagine aliens might find them interesting—or perhaps understand their otherworldly, ethereal song.

In the meantime, whales here on Earth remain mysterious. They live 99 percent of their lives underwater, far away from continuous contact with people and beyond most of our observational tools. We tend to think about them only when we glimpse them from the safety of a boat, or when they wash up along our shores. They also have an evolutionary past that is surprising and incompletely known. For instance, they haven't always been in the water. They descend from ancestors that lived on land, more than fifty million years ago. Since then, they transformed from four-limbed riverbank dwellers to ocean-going leviathans, in a chronicle we can read only from their fossil record, a puzzle of bone shards unevenly spread across the globe.

The little we have learned about whales leaves us unsatisfied because the scales of their lives and facts of their bodies are endlessly fascinating. They are the biggest animals on Earth, ever. Some can live more than twice as long as we do. Their migrations take them across entire oceans. Some whales pursue prey with a filter on the roof of their mouth, while others evolved the ability to navigate an abyss with sound. And then they speak to one another with impenetrable languages. All the while, in the short clip of our own history, we've moved from heedlessly hunting them to an awareness that they have culture, just as we do, and that our actions, both direct and indirect, put their fate in jeopardy.

A paleontologist is a good tour guide for what we know about whales, not just because their evolutionary history is profoundly interesting. It's because we, as paleontologists, are used to asking questions without having all of the facts. Sometimes we're losing facts: fossils removed from their medium lose clues of context; promising bonebeds are razed to make room for roadways; or bones lay misidentified in a museum drawer. When faced with these challenges, paleontologists turn to inference, drawing on many different lines of evidence to understand processes and causes that we cannot directly see or study—the same approach used by any detective, really. In other words, thinking like a detective is a useful approach to confront the mysteries posed by the past, present, and future of whales.

This book is not a synoptic, comprehensive account of every different species of whale—there are far too many whales to fit into anything shorter than an encyclopedia. Instead, this book presents a selective account, a kind of travelogue to chasing whales, both living and extinct. I describe my experiences from Antarctica to the deserts in Chile, to the tropical coastlines of Panama, to the waters off Iceland and Alaska, using a wide variety of devices and tools to study whales: suction-cupped tags that cling to their backs; knives to dissect skin and blubber from muscles and nerves; and hammers to scrape and whack away rock that obscures gleaming, fossilized bone.

The narratives in this book group into three general sections: past, present, and future. Broadly, I want to answer questions about where whales came from, how they live today, and what will happen to them on planet Earth in the age of humans (a new era that some scientists call the Anthropocene). But these stories don't cleanly fit into these three temporal silos. Instead, they build on one another and reciprocate because the ways that we need to think about whales require thinking about all the evidence at hand: unraveling the many mysteries of living whales requires a background about their evolutionary past, just as much as the surprises from the fossil record can clarify the meaningful facts about their lives today and into the future.

The first part of the book tells the chronicle of how whales went from walking on land to being entirely aquatic, relying on evidence from the fossil record showing what the earliest whales looked like. These fossils show us details that we couldn't otherwise know about the history of whales, and I explore exactly how we dig up these clues in the first place. Following fossil whale bones brought me to the Atacama Desert of Chile, where my colleagues and I puzzled over an ecological detective story with the discovery of Cerro Ballena, the world's richest fossil whale graveyard. How did this site come about, and what does it tell us about whales in geologic time?

The second part examines how and why whales became the biggest creatures ever in the history of life. The challenges of studying organisms as large as the largest species of whales means thinking about the limits of biology, and what exactly organisms at these superlative scales need to do, on a daily basis, to sustain their enormous sizes. While trying to connect muscle to bone at a whaling station, I share another serendipitous find: the discovery of an entirely new sensory organ in whales. What does an organ, lodged right at the tip of a whale's chin, mean for how, when, and why baleen whales evolved to become all-time giants?

Lastly, the third part explores the specter of the uncertain future that we share with whales on Anthropocene Earth. In the twentieth century alone, whaling in the open oceans killed more than three million whales, reducing many populations to shadows of their baseline abundances. Despite this decimation, no single species went extinct until the first decade of the twenty-first century. Since then, not a whistle or splash of the Yangtze river dolphin has been recorded, and responsibility for the extinction of this species can be placed squarely on our shoulders: we dammed the only river in which it lived. Other species, such as the vaquita, remain on the extinction watch list, numbering fewer than one dozen or two dozen individuals. But the news from the field isn't entirely dire: some whale species have rebounded from the brink, even expanding to new habitats as climate and oceans change. What can we imagine about our shared future with whales, drawing on their lives today and what we know about their evolutionary past?

Ultimately, the quest to understand whales is a human enterprise. This book is a story not just about knowing whales but also about the scientists who study them. The scientists described in these stories come from a variety of different disciplines, ranging from cell biology and acoustics to stratigraphy and parachute physics. Some are historical but very much knowable through their writings, their specimen collections, and the intellectual questions that they asked. One of the great privileges of my professional life is the opportunity to work at the Smithsonian, which has afforded me not only the latitude to undertake this pursuit but also firsthand access to some of the world's largest and most important collections of material evidence, be it specimens, scientific journals, or unpublished field notes. Every day, I think about the many generations of scientists before me who handled this same evidence, scratching away at the very same questions, while constrained by the circumstances of their times. My hope is that this book says as much about the inner lives of scientists as it does about whales.

PART I
PAST

1.

HOW TO KNOW A WHALE

I sat transfixed by a sea littered with a million fragments of ice, all rising and falling in time with the slow roll of the waves. We had spent the morning looking for humpback whales in Wilhelmina Bay, threading our rubber boat between gargantuan icebergs that were tall and sharp, like overturned cathedrals. Now we stopped, cut the engine, and listened in the utter stillness for the lush, sonorous breath of an eighty-thousand-pound whale coming to the water's surface. That sound would be our cue to close in. We had come to the end of the Earth to place a removable tag on the back of one of these massive, oceangoing mammals, but we took nothing for granted in Antarctica. As we sat waiting on the small open boat, I came to feel more and more vulnerable, a speck floating in a sea of shattered ice. "Don't fall in," my longtime collaborator and friend Ari Friedlaender deadpanned.

I struggled to remember how long we had been away from the *Ortelius*, our much larger oceanic vessel with its ice-hardened steel hull. In every direction, we were enclosed by a landscape of nunataks, jagged spires of rock that pierced the creamy tops of surrounding glaciers. Where the glaciers met the sea, they ended in sheer, icy cliffs towering over the bay.

9

Without a human structure for scale, these landforms seemed both near and far at the same time. This otherworldly scene of ice, water, rock, and light warped my sight lines, bending my sense of distance and the passage of time.

If you hold your fist with your left thumb out, your thumb is the western Antarctic Peninsula; your fist, the Antarctic continent's outline. The Gerlache Strait is part of a long stretch of inner passageway along the outer side of Antarctica's left thumb, and Wilhelmina Bay cuts a rough cul-de-sac off the Gerlache. The Gerlache is a hot spot for whales, seals, penguins, and other seabirds, and Wilhelmina Bay is the bull's-eye. All come here to hunt for krill, small crustaceans that form the centerpiece of Antarctic ocean food webs. Consider your hand again: individual krill are about the length of your thumb, but whales pursue them because they explode into great aggregations, or swarms, during the Antarctic summer. With the right mixture of sunlight and nutrient-rich water, dense clouds of krill form a sort of superorganism that can stretch for miles and concentrate in hundreds of individuals per cubic foot. By some measures, there is more biomass of krill than of any other animal on the planet. Calorie-rich swarms of them lurked somewhere, not far, just under our boat.

Where there are krill in sufficient quantities, there will be whales, but the fundamental problem with studying whales is that we almost never see them, except when they come to the surface to breathe or when we dive, in our own limited fashion, in search of them. Whales are inherently enigmatic creatures because the parameters of their lives defy many of our tools to

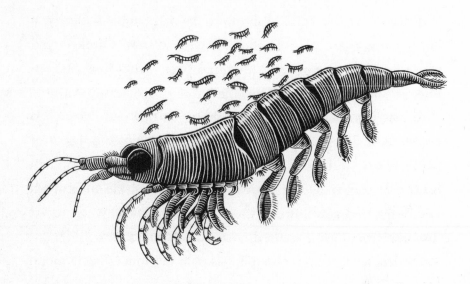

measure them: they travel over spans of whole oceans, dive to depths where light does not reach, and live for human lifetimes—and even longer.

In Wilhelmina Bay our goal was to attach a sleek plastic tag to the back of a whale to record audio, video, depth, changes in the whale's speed, and even pitch, yaw, and roll. Our tags would provide crucial context for how humpbacks interact with their environment by relaying, in a time-stamped way, how they feed on krill. Ari and his colleagues have tagged and tracked whales along the Antarctic Peninsula for nearly two decades, charting their movements against the backdrop of changes in krill-patch density, water temperature, daylight, and other variables. As climate change warms the poles faster than the rest of the planet, every year counts.

I sat on the gunnel of the boat while Ari scanned from the bow. We were several days into a multiweek expedition with high hopes to tag as many humpbacks as possible—ideally, a pod feeding together—but thus far we had been skunked. Ari stood rigid like a figurehead, holding a twenty-foot-long carbon-fiber pole in folded arms. The pole flexed in synchrony with the swell, the teardrop-shaped tag bobbing at one end. I watched clouds shift slowly overhead, mirroring the dappled quicksilver of the water, and wondered whether any other place on Earth could feel as alien as Antarctica. Then a loud gurgle interrupted my daydreaming, followed by a trumpeting blast of water vapor bursting from flaring, paired nostrils. A whale's blow.

We knew to expect more plumes of water immediately thereafter. A pod of whales synchronize when they surface to breathe—sometimes nearly simultaneously or split seconds apart. They usually breathe a few times in a row, in quick succession, before resubmerging—unless they're asleep or truly exhausted, whales tend to act like surfacing is a nuisance and they'd rather be deep underwater. Their tight coordination in breathing likely has a lot to do with maximizing time spent below the surface engaged in the cooperative tasks of finding food and avoiding predators. Some species travel or hunt in pods that are tightly knit family genealogies, while others, such as the humpbacks before us, form short-lived associations, seemingly a matter of happenstance.

"Oh, that's right," Ari called out. The vapor from the blow lingered in the cold air. Ari pointed to a small patch of water a dozen yards from the boat, perfectly calm against the waves at the surface. This patch, called a flukeprint, betrayed the whale's movement, unseen deep below our tiny boat. The single fluke-print bloomed into several, each the size of our boat, rising up from the depths, whirling and spreading into the smooth geometry of a lily pad. We were right. "He's got buddies," Ari said. Without the aid of an echo sounder—which would also tip off the whales about our location—we used the ephemeral patterns on the surface to read their path.

We started up the engine, throttling ahead slightly to a spot beyond the last flukeprint. Within seconds, right on cue, a pair of enormous nostrils bubbled at the surface, releasing a thundering tone and then a spray that carried past us as we kept up with the whale. A dorsal fin surfaced, and a second and third blow exploded nearby. "Pull up behind this last one. We have about three more breaths until they go down," Ari shouted.

We trailed the laggard of the group, maneuvering the boat into position. Ari lowered himself across the bow and held the pole against his torso, extending the tip with the tag just ahead of the dorsal fin as we motored close to the behemoths moving yards away. Then, in a decisive motion, Ari launched the tip of the pole toward the whale's back, where the tag's suction cups hit the skin with a satisfying *thwack*. The whale rolled beneath the surface as we pulled back to pause and wait for it to return. We spotted its sleek, shiny back as it arose again, marked with a neon tag, and we cheered. The whale took one last breath before it raised its monstrously wide tail flukes out of the water and slipped down into emerald darkness with the others. Ari radioed back to the *Ortelius*. "Taaaaag on," he said with a hint of swagger as he grinned at me.

Tag on

The whole rodeo of tagging is a bit like sticking a smartphone on the back of a whale, complete with the logistics of getting close enough to a forty-ton mammal in the first place. Just as your smartphone can record movies, track where you go, and automatically rotate images, the same technology—miniaturized and cheap, combining video, GPS, and accelerometers all in one device—has fueled a revolution in understanding how animals move throughout their world. Scientists call this new way of recording organismal movement biologging, and it has captured the interest of ecologists, behavioral biologists, and anatomists, all interested in knowing the details of how animals move through space and over time. Biologging has been especially important for revealing the daily, monthly, and even annual meanderings of animals that are extremely difficult to study. Stick a tag on a penguin, a sea turtle, or a whale, and there's a chance to know how it swims, what it eats, and everything else it does whenever you're not around to watch it—which is most of the time, for animals at sea.

The logistics of studying whales places them in a realm truly apart from every other large mammal on land or at sea. To know anything about them in the wild takes time on a boat, sticking a tag on their back, sliding a camera underwater, or spying over-head with a drone—if you're lucky enough to come upon them in the first place. Biologging is helping us overcome this chal-lenge by giving us a remote view into their lives, an extension of our senses more intimate and sometimes more detailed than any telephoto lens. In the case of humpbacks, tag data have revealed how these gulp-feeding whales lunge at large schools of krill and other prey, oftentimes in coordinated attacks. It's a form of pack hunting, which might seem strange for a species celebrated as gentle giants. But baleen whales are serious predators, not like grazing cows but more like wolves or lions, pursuing their quarry with strategy and efficacy. Don't be fooled by their lack of teeth, or just because krill don't bleat in terror.

Hours later the *Ortelius* prowled silently through Wilhelmina Bay, with a pair of massive spotlights leading the way in search of icebergs in our path. Outside on the bow, I watched heavy snowflakes drift through the cones of light as Ari unfolded the metal radio antenna that would lead us back to our tag. The tags we were using had to be re-collected to yield their data; we would have to find and physically pluck them out of the water, provided that they'd already been knocked off the whales' backs. By design, they can last minutes, hours, or even days from the force of suction alone, before being scraped, being bumped, or falling off. Its buoyant neon housing would keep the whole device floating on the surface until we triangulated its position.

Over the course of his career, Ari has probably tagged more species of whales than anyone, in all circumstances and in every ocean. Beyond friendship, mutual colleagues, shared ambitions, and wry humor, we were working together in Antarctica to build a bridge between our disciplines—paleontology for me, ecology for him—because questions about how whales evolved to become masters of ocean ecosystems over fifty-odd million years need to be grounded in the facts of what whales do today. Bridging gaps between disciplines sometimes necessitates spending time side by side. All the better if it's in the field.

The metal prongs that Ari assembled looked like an elaborate set of rabbit ears from an old television set. He plugged the antenna into a small receiver with a speaker, and after a few moments, we heard a series of intermittent beeps. "That gap between the beeps tells us that the whale is sleeping, rising up to the surface to breathe, and then sinking back down." Ari smiled. "Just dozing, belly full of krill. Not a bad way to spend a Saturday night." We would need to come back later and listen again for our tag until it floated freely, beeping uninterrupted.

Most large baleen whale species alive today belong to the rorqual family, which feed on krill and other small prey by lunging underwater. They comprise the more familiar members of the cetacean bestiary, including humpbacks, blue whales, fin whales, and minke whales. Rorquals are also the most massive species of vertebrates ever to have evolved on the planet—far heavier than the largest dinosaurs. Even the smallest rorquals, minke whales, can weigh ten tons as adults, about twice as much as an adult bull African elephant. Rorquals are easy to distinguish from any other baleen whale, such as a gray whale or a bowhead whale: look for the long, corrugated throat pouch that runs from their chin to their belly button. (And yes, whales have belly buttons, just like you and me.) The features that make rorquals so obviously different from other baleen whales also play a critical role in how they feed.

Across whole ocean basins, individual whales find their food using probability, heading for feeding grounds burned into memory from a lifetime of migration. Rorquals travel routes that span hemispheres over the seasons; an individual whale might migrate from the tropics in the winter in search of mates and to bear young, then to the poles during the summer to forage under constant sunlight. Baleen whales still retain olfactory lobes, unlike their toothed cousins, such as killer whales and dolphins, which have lost them. Baleen whales might smell some aspect of their prey at the water's surface, and it is possible that this mechanism could refine their search once on the scene. Originally their sense of smell evolved for transmission through air, not water; we know little beyond the basics about this sense in whales. Somehow, whales manage to be in the right place at the right time to feed. And what's clear from biologging is that once in the right place, baleen whales spot the prey patches from below, probably approaching them by sight. Lacking the echolocation of their toothed relatives, vision is likely the dominant sense for baleen whales at short range.

With prey in range, a rorqual accelerates, fluking at top speed, and begins the amazing process of a lunge. Surging from below, it opens its mouth only seconds before it arrives at a patch of krill or school of fish, which may be as big as or bigger than the entire whale. When it lowers its jaws, the rorqual exposes its mouth immediately to a rush of water that pushes its tongue backward, through the floor of its mouth, into its throat pouch. In mere seconds, the accordion-like grooves of its throat pop out like a parachute. After engulfing the prey-laden water, the whale slows, almost to a halt, pouch distended and looking bloated, nothing like its airfoil-shaped profile from moments prior. Over the next minute, it slowly expels water out of its mouth through a sieve of baleen, until its throat pouch returns to its original form, the prey swallowed. For their part, krill and fish deploy collective defensive behavior by dispersing to try to escape the oncoming maw of death. In the end, a successful whale takes a bite out of a much bigger, more diffuse and dynamic superorganism.

Lunge feeding has been described as one of the largest biomechanical events on the planet, and it's not hard to imagine why when you consider that an adult blue whale engulfs a volume of water the size of a large living room in a matter of seconds. Tags on humpbacks in other parts of Antarctica show how they sometimes feed close to the seafloor in pairs, swimming alongside each other as they scrape the bottom with their protruding chins in mirrored unison. Tags have also shown us that rorquals are right- or left-handed, just like us, favoring either a dextral or sinistral direction when they roll their bodies to feed.

The more scientists tag whales, the more it's apparent that there's still much that we don't know. It turns out that blue whales have a behavior where they spin 360 degrees underwater in a pirouette before they lunge, probably to line up their mouths precisely with a patch of krill. Other lightweight tags, launched with barbs that cling more deeply beneath the skin on the dorsal fin, have tracked the movement of Antarctic minke whales migrating over eight thousand miles of open ocean, from the Antarctic Peninsula to subtropical waters. These tags upload data directly to satellites whenever the whale surfaces, over the course of weeks to months, before eventually falling out. These tags are also especially useful for species that are rarely seen, such as beaked whales. Satellite-linked dive tags deployed on Cuvier's beaked whales revealed, in a precise way, the astonishing extremes of their foraging dives for squid and fish—over 137.5 minutes of breath holding, 2,992 meters deep— data that set new dive records for a mammal. If the idea of holding your breath for over two hours doesn't alarm you, imagine doing it while chasing your dinner to a depth of nearly two miles.

Tag data combined with tissue samples taken from biopsy darts tell us that these humpback whales feeding in the western Antarctic Peninsula are merely seasonal visitors for the austral summer. By early fall, they depart the icy bays, cross the great Circum-Antarctic Current that rings the seventh continent, and undertake various paths over thousands of miles to arrive at temperate latitudes. At Wilhelmina Bay the overwhelming majority of humpbacks return to the low latitudes of the Pacific coasts of Costa Rica and Panama to mate and give birth before returning to the Southern Ocean for the next austral summer to feed.

We eventually retrieved the tag, along with its data, and continued to Cuverville Island, on the other side of Wilhelmina Bay. As the *Ortelius* maneuvered out of the Gerlache and toward the island, I watched from the ship's stern as we passed icebergs more massive than any I had yet seen. Their fragmented sides, a hundred feet high, were etched in luminous, milky blues and grays. They held light from sea to sky, glowing in unearthly ways, as if they could not have been formed on this planet. And of course they were mostly sheathed underwater, which was a bit of an ominous thought; the *Ortelius* kept a careful distance. But the incomprehensibility of something as overwhelming as an iceberg is belied by its transience: even the largest ones, platforms the size of cityscapes, will eventually shed their layers of ice, annealed over hundreds of thousands of years, and become part of the sea.

Scattered around the peninsula are several islands like the one we approached, islands that served as barely inhabitable platforms for whaling operations in the early and midtwentieth century. Today the only remnants of human civilization are occasional concrete pylons with bronze plaques identifying the area as an open-air heritage site, and leftover whale bones. After we hauled our rubber boat up on the rocks, I walked toward the spoil piles of green-stained and weathered whale bones, strewn like spare lumber at a construction site.

Reading whale bones is what I do, although sometimes I feel like the bones find me. I've spent so much time searching for them, cataloging them, and puzzling over them that my brain immediately recognizes even the slightest curve or weft of bone. Whale bones tend to be relatively large, so finding them is often largely a matter of making sure that you're in the right neighborhood—it shouldn't have been much of a surprise, especially on the grounds of an abandoned whaling station. On the island I mentally inventoried the first assemblage I encountered, as I dodged foot-tall gentoo penguins scrambling at my feet: ribs, parts of shoulder blades, arm bones, and fragments of crania. They clearly belonged to rorqual whales, about the size of humpbacks, or possibly even fin whales. Some of the more intact vertebrae were artfully balanced upright on the shoreline, probably posed by Antarctic tourists, passing through the peninsula by the thousands in the austral summer and looking for a perfect photograph.

If these bones belonged to humpback whales, it would not be surprising, given the abundance of this species out around Antarctica today. It's likely that some of the whales that we tagged were descendants of these individuals, belonging to the same genetic lineage. But history tells us that if you turned back the clock a century, humpbacks probably wouldn't have been the only ones here: blue and fin whales would have numbered in the hundreds, if not thousands; minke whales, beaked whales, and even Southern right whales would also have been part of the community. Ari has seen only one right whale out of the thousands of whales that he's observed over fifteen years in the area. Southern right whales have barely recovered from two hundred years of whaling, and we know little about where they go besides their winter breeding grounds along protected coastlines of Australia, New Zealand, Patagonia, and South Africa.

It's not just right whales that vanished. There's no memory or record of just how many of any kind of whale there was in the Southern Ocean, in terms of their abundance, before twentieth-century whaling killed over two million in the Southern Hemisphere alone. However, as whale populations in this part of the world slowly recover from this devastation, we're beginning to see what that past world might have looked like. On an expedition in 2009, Ari and his colleagues documented an extraordinary aggregation of over three hundred humpbacks in Wilhelmina Bay, the largest density of baleen whales ever recorded. "There is no external limit on these whales because there is just so much krill. They literally cannot eat enough before they need to leave," Ari reflected. "That incredible resource base means that it's just a matter of recovery time for whales— and I think what we saw in the bay that year was a glimpse of what their world was once like, before whaling." On the whole, humpbacks have recovered to only about 70 percent of their prewhaling numbers in the Southern Ocean, although along the peninsula their population size has nearly returned to the best estimates of prewhaling levels at the start of the twentieth century.

I paused on a guano-free ledge to record a few observations about the bones' weathering and their measurements in my field notes. To the southwest, the sky churned in a dark gray, portending wind and snow, and I felt a chill creep into my damp toes and fingertips. I pulled off my gloves and reached for a disposable hand warmer in my jacket pocket. Lodged in a mess of receipts and lozenge wrappers was a note my son had left me on the kitchen counter back home:

Im gona mis you
wen you go to
anaredica.

The night before I left my home in Maryland we traced the expedition route on a plastic globe. When he wanted to know how far away eight thousand miles was in inches, I didn't tell him the answer that I wanted to, which was "Too far." I reassured him that the passage was safe and that we would stay warm. "I'll think about you when we drink hot cocoa," I offered, dressing up my own concerns with a good smile.

As we pulled away from Cuverville Island to return to the *Ortelius*, the swirling clouds began to send down flurries, covering us in thick, wet snow. The boat bumped hard against the waves, and we saw humpbacks surfacing far off in the distance, the wind pushing their blows quickly behind them. The sight of those living, breathing, feeding whales in the same view as the island with beach-cast bones made me feel as though I could see the present and past simultaneously, each telling us facts that the other vantage could not. The bones on Cuverville Island and Ari's tagging work in the Gerlache were each a unique window into the story of humpback whales in the Antarctic, though these views were terribly incomplete: the past represented by mere bones crumbling on remote shores, what we know today limited to a few hours' or days' worth of data collected by a hitchhiking recorder on whales' backs.

Scientists tend to operate within intellectual silos because of the years of training and study that it takes to know about any single part of the world. But the best questions in science arise at the edges. Ari and I both want to know how, when, and why baleen whales evolved to become giants of the ocean—Ari wants to know more about their ecological dominance today, and I want to know what happened to them across geologic time. The answer to the basic question about the origin of whale gigantism requires pulling data and insights from multiple scientific disciplines, which is another way of saying that we need the perspectives of different kinds of science—and scientists— to untangle the monstrous challenges of the nearly inaccessible lives of whales. That's why a paleontologist like me was on a boat tagging whales at the end of the Earth: I needed a front-row seat to know exactly what we can hope to know from a tag. But answering the questions that most captivate me about whales requires more than just a single tag. It means wrapping my arms around museum specimens, handling microscope slides, paging through century-old scientific literature, and wading knee-deep in carcasses.

The wind sapped the last warmth from my already-wet gloves and whipped through openings around my hood as I held tight to the ropes on the gunnels. The first scientists to visit this place, over a hundred years ago, didn't have the luxury of disposable hand warmers. They suffered more brutally than we can really imagine, with less certainty of safe return. In these narrow margins they must have wrestled with the tension that overcomes scientists in the field: the desire to apprehend something almost unknowable against the tolls of living a world away from civilization. I patted my son's note, folded safely inside my jacket pocket. Hot cocoa sounded just right.

2.

MAMMALS LIKE NO OTHER

I was never a whale hugger. I didn't fall asleep snuggling stuffed whales or decorate my room with posters of humpbacks suspended in prismatic light. Like most children, I went through phases of intense study: sharks, Egyptology, cryptozoology, and paleontology. The curriculum was loosely inspired by my small curio cabinet crammed with a bric-a-brac collection of gifts and found treasures: abalone shells from my parents' friends in California and fluorite from a great-aunt in New Mexico sat next to trilobites and fossil ferns that I had collected on family trips to Tennessee and Nova Scotia (good fossils being hard to come by on the island of Montreal). My collection was a tangible means to escape, across geography and time, as I read ravenously about dinosaurs, mammoths, and whales under the tacit encouragement of my parents, professors who recognized this type of aimless curiosity.

During one of my immersive phases, I came across a distribution map that showed the location of whale species around the world. With my finger I traced the range of blue whales, the largest of all whales, as it went right up the St. Lawrence River, which bordered my neighborhood. I wondered about my chances of seeing a blue whale casually surfacing in the distance near my house. The thought of a local blue whale was a reverie that often arose in my mind as a kid, although it took two decades for me to return to it in earnest, as a scientist.

Some branches on the tree of life become quite personal, for reasons that are difficult to explain. We seek reflections of parts of ourselves in beings seemingly close to us—the disdain of a house cat or the perseverance of a tortoise—but in the end these species are distinctly other, refashioned by evolution and eons of time away from our shared ancestry. Those differences are accentuated to the furthest degree in whales; they seem mostly other—otherworldly, really—and that makes them both fascinating and enigmatic. They embody an incongruity that is vexing because they betray their mammalian heritage in so much of what they do, yet they look and live so far apart from us. Their size, power, and intelligence in the water are astonishing because they're unparalleled, yet whales are benign and pose no threat to our lives. They are almost a human dream of alien life: approachable, sophisticated, and unscrutable.

I don't malign whale huggers and dolphin lovers, even if I wrinkle my nose at the rhapsodic celebrations of armchair experts. Yes, whales and their lives are superlative, foreign, and well worth epic prose. But their amazing qualities are just starting points for me, as a scientist. Whales aren't my destination: they are the gateway to a journey of discovery, across oceans and through time. I study whales because they tell me about inaccessible worlds, scales of experience that I can't feel, and because the architecture of their bodies shows how evolution works. By rock pick, knife blade, or X-ray, I seek the corporeal evidence they provide—their fossils, their soft parts, or their bones—as a tangible way to anchor questions that surpass the bounds of our own lives. Whales have a past that reaches into Deep Time, over millions of years, which is important because some features of these past worlds, such as sea level rise and the acidification of ocean water, will return in our near-future one. We need that context to know what will happen to whales on planet Earth in the age of humans.

Whales are so very unlike the furry, sharp-eyed, tail-wagging, baby-nuzzling animals we think of when it comes to our mammalian relatives. First off, whales are among the few mammals that live their entire lives in the water. The only fur to be found on their bodies is the hairs that dot their beaks at birth. Although whales possess the same individual finger bones that you and I do, their phalanges are flattened, wrapped together in a mitt of flesh, and streamlined into bladelike wings, no hooves or claws to mar their perfect hydrofoils. Hind limbs exist only as relics in a handful of species, bony remnants tucked deep within muscle and blubber. A whale's backbone ends in a fleshy tail fluke, like a shark's; but unlike a shark or even a fish, whales swim by flexing their backbone up and down, not side to side. In short, they look nothing like squirrels or monkeys or tigers, but whales still breathe air, give birth, nurse their young, and keep company with one another over their lifetimes.

Fossils tell us the earliest whales were more obviously, visibly mammalian. The first whales had four legs, a nose at the tip of their snout, and maybe even fur (up for some debate among paleontologists, as fur doesn't readily fossilize). They had sharp, bladelike teeth and lived in habitats that ranged from wood-lands with streams to river deltas, occasionally feeding in the brackish waters of warm, shallow equatorial coasts. The oldest fossils of these land-dwelling, four-legged ur-whales come from rock sequences around about fifty million to forty million years old in the mountain ranges of Pakistan and India. At the time, the Indian subcontinent had not yet collided with Asia and sat in the middle of the forerunner to the Mediterranean Sea, called the Tethys sea, which split the Old World at the equator.

The skeletons of most of these first whales were the size of a large domestic dog. Because they lived on land, you won't find the flattened arm and finger bones we see in whales today—instead their limb bones are round and weight bearing, and their hands and feet end in elegant, delicate phalanges. Their tail, as far as we can infer from the available bones, did not end in a fluke. Their Latin names give some clues about their prov-enance or what makes them special. *Pakicetus*, for example, originates from an area that is now Pakistan, but was once an island archipelago where early whales climbed in and out of streams. *Ambulocetus*, a low-slung early whale with body and skull proportions like a crocodile, has a name that translates as "ambulatory, or walking, whale." *Maiacetus*, one of the rare early whales for which we have a near-complete skeleton, earned its name from fetal bones preserved near the abdominal cavity of the original specimen—the mother whale. Today's whales all give birth tail first; the rear-facing position of the fossilized fetal *Maiacetus* showed that whales at this evolutionary stage still gave birth on land, headfirst.

The combination of four legs, phalanges, and cusped teeth is

Pakicetus, *a land dweller, swims in an Eocene streambed.*

found in no whale alive today. What made these ancient creatures whales in the first place is subtle, lodged deep in their
skeletons. That's a good thing for us because these hard parts
stand a chance of being preserved over tens of thousands of millennia. One of the most important features is the involucrum, a
fan-shaped surface on the outer ear bone, rolled like a tiny
conch shell. *Pakicetus* has an involucrum, as does every other
branch on the whale family tree subsequent to it. The involucrum is one key trait, along with small clues in the inner ear and
braincase, that the earliest whales share exclusively with today's
whales and no other mammals. In other words, it's a feature
that makes them whales and not something else. It's unclear
whether the trait gave *Pakicetus* an advantage for hearing on
land, but later lineages of early whales co-opted it to hear directionally underwater, using a connection between the outer ear

bones and the jawbones. Tens of millions of years later, the involucrum (and underwater hearing) persists in today's whales, from porpoises to blue whales.

Fifty million years of whale evolution can be split into two major but unequal phases. The first deals with the transition of whales from land to sea in less than ten million years; the earliest land-dwelling whales all belong in that first phase—even at their most aquatic, they still retained hind limbs that could have supported their weight on land. The second phase covers everything that happened once whales evolved fully aquatic lives, for the remaining forty or so million years, until today. Throughout both of these phases, extinction dominates as a constant background theme because, as with the vast majority of animal lineages on the planet, most all the whale species that ever evolved are now extinct. While they are the most diverse marine mammal group today, numbering over eighty species, the fossil record documents over six hundred whale species that no longer exist.

The first phase of whale evolution is fundamentally about transformation: the tinkering and repurposing of structures from an ancestral state (originally for use on land) to a new one, in aquatic life. Transformation requires an initial state, and some starting points in evolution can be difficult to discern. For example, hearing, sight, smell, and taste are all senses that evolved for nearly 300 million years on land before the first ancestors of whales took to the sea. While it's convenient to think of the reshaping of hands into flippers in whales as an undoing, that's a mistake: whales didn't undo 300 million years of terrestrial modifications. They did not, for example, recover gills. Instead, the story is far more interesting. Whales worked with what their ancestors had as land animals, modifying many anatomical and physiological structures for a new use rather than some phantasmagoric evolutionary reversal.

The second phase, after whales got back in the water, encompasses any whale lineage obliged to spend its life exclusively in the water; this phase also spans all of the consequences that arise from that constraint. You can think of evolutionary innovation as a hack on constraint. In other words, novelty in evolution is the appearance of a totally new structure, such as baleen, that confers not just a slight advantage to those who possess and inherit it but shifts their descendants into a completely new dimension of adaptation. The second phase of whale evolution, when innovations such as filter feeding and echolocation appear and fuel the diversification of today's whales, stretches in time from the first aquatic whales, about forty million years ago, to the present day, including all living cetaceans, along with hundreds of extinct forms in between.

In the past 250 million years, many backboned animals converted from living in terrestrial ecosystems to living in oceanic ones. The first wave happened throughout the time of the dinosaurs, when many different reptile lineages invaded ocean ecosystems from 250 million to 66 million years ago. Since the mass extinction at the end of the Cretaceous, the ecologically dominant ocean invaders have been mammals—including everything from whales to sea otters—although penguins and Galápagos marine iguanas are also more recent reentrants. All of today's marine mammal lineages are distantly related to one another, whether it's a whale, a sea otter, a seal, a sea cow, or a polar bear (yes, technically polar bears too, which eat seals and hop across ice-covered seas).

What makes early whale evolution so important is that the completeness of the fossil record from the early stages—*Pakicetus*, *Ambulocetus*, *Maiacetus*, and all others like them during that first phase—is unmatched by any other group in the fossil record. We simply don't have the range of fossils showing the specific anatomical transformations from land to sea for any other mammal or reptile the way we do with whale origins.

Even so, the evidence for whale origins has only recently been uncovered. Until about forty years ago, we had no idea what the hind limbs of the earliest whales in the first evolutionary phase really looked like. The discovery of *Pakicetus* in 1981 gave us mostly bones from the neck up—paleontologists discovered a small W-shaped braincase exhibiting, among other features, an involucrum, but it otherwise looked like any other land mammal's. They found the skull—pinched and delicate, like a handheld vase—in river deposits, and concluded that the earliest whales lived some part of their life on land. Without more of a skeleton, at the time they could only speculate about what these whales looked like from the neck down.

In 1994 the discovery of *Ambulocetus* clarified this picture, showing that the earliest whales had weight-bearing fore and hind limbs, with separate phalanges perhaps connected in life by webbing. Relatively large feet in *Ambulocetus* were a clue about its swimming style, which likely involved flexing its spinal column along with its broad feet, in one motion. Mechanically this style is somewhere between paddling with hands and feet (using drag for forward motion) and employing a hydrofoil, as modern whales do with their tail fluke (using lift, instead of drag). Our pelvis is rigidly connected to our backbone, whereas in *Maiacetus*, the pelvis was only partially connected to the backbone, permitting a lot of flexibility for the whole spinal column to undulate up and down. The shape of a few tail vertebrae can reveal a lot about locomotion—in *Ambulocetus* the fact that the tail vertebrae are longer than they are tall tells us that these early whales had long, thickened tails, although we still don't have enough bones to know what direction these powerful tails might have moved.

Ambulocetus still didn't provide enough evidence to help answer the big questions about whale origins: Where did they fit into the mammalian family tree? Who are their closest relatives? By the 1990s, DNA studies had shown that hippos are the closest living relatives to whales. Hippos and other even-toed hoofed mammals, such as cows, deer, and pigs, are seemingly unlikely relatives, until you look at their stomachs. Even anatomists in the nineteenth century knew that living whales had multichambered stomachs like these ungulates, pointing to a possible evolutionary relationship. Paleontologists, however, had other extinct fossil mammals in the running for whale's closest relatives: mesonychids, which had strikingly similar teeth and were wholly carnivorous, as whales are today, but left no descendants. Without more skeletal material from four-legged whales, especially from their limbs, there was no way to parse the stories of DNA versus fossils for the deepest origins of whales.

Then, in 2001, two competing groups of paleontologists reported the same pivotal piece of evidence from different species of early whales: they each had discovered that the anklebone of ancient land-dwelling whales was exactly like those of living even-toed ungulates. This bone, called the astragalus, looks like two 35 millimeter film canisters taped together like a raft; in your hand it feels like some kind of board-game piece. Cows, goats, and camels all have it. Living whales don't because they have no feet, and the only traces of hind limbs are reduced to nubbins of bone next to free-floating pieces of their pelvis, wrapped deeply in their body walls—making fossil hind limbs in early whales the only source for this information. Mesonychids didn't have these double-pulley anklebones, which meant their tooth similarities with early whales were the result of convergent evolution—something that has happened frequently in mammal evolutionary history. The discovery that early whales had a so-called double-pulley astragalus confirmed the DNA findings: whales were just highly modified even-toed hoofed mammals, minus the hooves.

Since finding *Pakicetus*, paleontologists working in remote parts of Egypt, Pakistan, and India have discovered a rich variety of early land-dwelling whales that lived about fifty million to forty million years ago, toward the end of a geologic epoch called the Eocene. These early whales seem to have been experimenting with ecological modes that have parts both familiar and strange: *Ambulocetus* looked crocodile-like; *Maiacetus* more like a sea lion, which had not yet evolved; still other strange early whales such as *Remingtonocetus* were an amalgam of zoological categories, something like a long-snouted otter; and *Makaracetus*, named after a mythological South Asian creature that is half fish, half mammal, had a downturned snout, perhaps for eating clams. All of these early whales belonged to extinct branches at the base of the whale family tree; our expectation about what makes a whale is hindsight biased, based on how we see them today—a great challenge paleontologists face when trying to understand the biology of these extinct whale relatives.

Knowing how whales turned out makes our retelling of their evolutionary pathway a tidy, preordained story. It's easy to imagine *Pakicetus*, looking something like a lost dog dipping its toes in the water, followed then by intermediate stages of creatures each spending more time in the water: *Ambulocetus*, which could hear underwater and lunge at prey with its powerful limbs, like an ambush predator; followed by *Maiacetus*, whose pelvis was less strongly coupled to its spinal column, permitting the first kind of flexibility for tail-driven propulsion in whales. Fossils belonging to relatives of *Maiacetus* extend over a far greater geographic range than those of previous ur-whales, suggesting that this still-quadrupedal animal was seafaring, though it still returned to the shore to give birth, like sea lions today. In this view, *Maiacetus* represented the last of the earliest whales; all subsequent whales, in the second phase of the evolutionary chronicle, had no weight-bearing limbs and were totally separated from land.

The problem with this linear narrative is that we know the final result, which lets us pick and choose the likely path of least resistance toward the whales we recognize today. But evolution doesn't work like that: it makes no concession for the future; it's about what's good enough in the moment. Selection operates on what's available, sorting biological variation based on the demands of the immediate world. If you were somehow able to return to a late Eocene shoreline in the Tethys sea and happen upon the entire assemblage of early whales in one lineup—all of the early whales, four-legged and odd, scattered on the shoreline—you wouldn't be able guess the eventual winner of the evolutionary sweepstakes. In its own time and habitat, each early whale was as well adapted as any crocodile, sea lion, or otter living today. It's just that when we work with the fossil record, we're afforded a view of the very long run, and the relative successes and failures in any particular group over millions of years. The eventual winners of the evolutionary sweepstakes were early whales that completely severed their ties to land, becoming fully aquatic, eventually yielding descendants that filter feed and echolocate.

These first whales were merely semiaquatic mammals with specializations for life near the water to one degree or another. There was nothing predetermined about some of their descendants becoming fish-shaped leviathans many millions of years later. Retrospection, however, does cue us into specific features that show incremental transformations: shell-shaped ear bones being repurposed for underwater hearing; or the pelvis becoming unlinked from the backbone, allowing the whole back end to serve as a propulsion device. If you focus on tallying which species go extinct and which ones survive, you might lose sight of the important lessons about major evolutionary change told through bones over geologic time.

Of all the two-hundred-odd bones in a whale's body, skulls are probably the most important part to examine if you're interested in the big picture of whale evolution. Like the skull of any vertebrate animal, whale skulls past and present conveniently house the primary organs for taste, smell, sight, sound, and thought all in one unit. Skulls are thus rich sources of functional information about the lives of whales and their transformation over time—after all, these senses are tweaked, enhanced, or diminished when lineages undergo major ecological transitions, such as the one from land to sea, over the course of evolution. Despite their durability, skulls are challenging objects for study. Their individual bones interlock with one another in complex and hidden ways, with blind corners, overlapping parts, and delicate connections. Soft tissues such as the eyes and brain all rest across several bones, like fruit sitting in a bowl made of interconnected puzzle pieces. To make things even more interesting, whale skulls are not only intricate but big. I've stared at whale skulls long enough that they feel familiar to me, but I always have to remember that whale skulls are, in very clear ways, unlike those belonging to any other mammal.

Take the skull of a bottlenose dolphin, which rests comfortably on a desk but would require two hands to move carefully. It consists of two basic parts: a paddle-shaped beak, formed of elongate bones with rows of teeth like pencil tips; and a cranium of layered bones that cover a bowling ball–shaped braincase. About those teeth: You won't find the traditional lineup of incisors, canines, premolars, and molars that mammals usually possess. At some point in their evolutionary history, toothed whales gave up chewing for merely seizing their prey with a snap of their jaws and then swallowing it whole. A bottlenose dolphin may flash what looks like a welcoming toothy grin, but I wouldn't put my hands anywhere near it.

Moving from the beak to the rest of the skull, the next-most-obvious feature is the orbit, the bone roofing where the eye would be, like a heavy eyebrow, still very much like that of other mammals. But behind the orbits, differences begin to accumulate. First, there's the aperture that leads to nostrils, or the blowhole. You can peer down the curved passageway formed by these nostril bones to the underside of the skull, where you'll see an origami construction of delicate, folded bones with paper-thin edges. The bones leading to the blowhole are actually behind where the eyes would be located, the complete opposite of any other mammal, where nostrils are located at the tip of the snout. If your nostrils were positioned like those of a dolphin, you'd blow your nose from the top of your forehead.

Pakicetus, *Maiacetus*, and *Remingtonocetus* had nostrils toward the tip of their snout, and these structures slowly migrated backward in other stages of fossil whales, up to the bottlenose dolphins that we see today, with nostrils displaced well behind the eyes. Interestingly, whales aren't alone in nostril migration: sea cows and manatees, also full-time aquatic mammals, have nostrils positioned high on the skull (though not behind the eyes), whereas their early fossil relatives have nostrils positioned more forward. This parallel migration of the nostrils lets fully aquatic mammals, such as whales and sea cows, orient their body in a more energy-saving horizontal position in the water, as opposed to doggy-paddling with their nose out of the water. But swimming horizontally is just part of the reason for the strangeness of the bottlenose dolphin skull before us.

When viewed from the side, the top of a bottlenose dolphin's skull is shaped like a scoop—if you were a dolphin, imagine having a skull with a dishlike forehead, right above your eyebrows. In life, for a dolphin this cavity contains a cone of fat called the melon, which gives toothed whales a domelike forehead. Tucked behind the melon, and underneath the blowhole, are empty sacks: air sinuses underlain by muscles and sealed with an organ that looks like a pair of lips. When these lips buzz like a trumpeter's, they generate sound, which bounces around the inside of the head and then gets focused by the melon into a discrete path out of the head, as a high-frequency sound beam, like an acoustic beam emanating from a searchlight strapped to the dolphin's forehead.

By coordinating this process across a specialized set of anatomical parts to generate sound, toothed whales create a form of biological sonar—or echolocate—to see their underwater world in sound. All toothed whale species alive today echolocate, whether they are sperm whales, beaked whales, river dolphins, porpoises, or true dolphins. It's how they wayfind, or hunt, in murky rivers or at ocean depth, sometimes a mile deep, with no light. Echolocation itself has evolved only a handful of times in other vertebrates; toothed whales are the only animals that do it underwater.

Making high-frequency sound, however, is just one part of echolocation; after sound bounces off an object, it produces an echo that the animal needs to hear. (The classic dolphin chirps and squeaks are how they vocalize to communicate with one another at lower frequencies.) We can't hear directionally underwater, but whales can because their ear bones float inside hollow pockets. In life, the ear bones of a dolphin hang in a sinus cavity full of spongy tissue, which acoustically isolates each ear, letting the brain detect small differences in the arrival time of sound between the right and left ear, helping to pinpoint the source in three dimensions. Whales have had the ability to hear like this since the time of *Ambulocetus*.

But how does sound get to the ear bones, especially when whales don't have external ears? Fats are conductive of sound, and along with the melon, toothed whales have large, fat bodies that fit hand in glove into hollows in the backs of their jaws and then branch out, backward, into lobes that directly connect to the ear bones. Akin to the way our ear canals funnel sound, these fat bodies provide a pathway for sound to reach the ears, although researchers still debate whether there aren't other acoustic pathways in the head—testing the anatomical basis for echolocation requires clever experimental work, and captive animals are not easily procured. This hallmark sense in toothed whales was poorly understood until the late 1950s, when scientists, supported by strong interest from the U.S. Navy, resolved the enigmatic echolocation abilities of toothed whales in a simple way: after covering a captive dolphin's eyes with suction cups, they discovered that the animal was still able to navigate a maze to find a small object.

We still know little about how echolocation in toothed whales truly works, especially in wild, free-ranging animals. It surpasses even the best military technology, and we're just starting to understand the basics. There is marked variation, from species to species, in terms of the arrangement of air sacs, melon shape, and even ear bones. What any one of those differences means for sound frequency and how whales perceive it remains a broad mystery, hopefully the work of future dissertations. We can say, however, that a quick tour of a dolphin skull shows us how evolution refashions or modifies existing parts—like nostrils displaced behind the level of the eyes—while other times it generates completely new structures, such as a biosonar apparatus on the forehead. How evolutionary novelty happens remains one of the most important unanswered questions in biology today.

Transformation and novelty are recurring themes in whale evolution as much as they are for any group in the tree of life, and it is sometimes difficult to parse the distinction between them. Consider how the feathers that most birds use today for flight are elaborations—transformed many, many times over—from basic scales that once covered their dinosaur ancestors. Or how turtles, early in their evolutionary history, acquired shoulder blades inside their rib cage; turtles' ribs later fused with other bones so that every turtle thereafter, for the past 200 million years, has had a shell with its shoulders tucked firmly inside. In both of these cases, evolutionary novelty does come from somewhere in the body—it's an extreme flavor of transformation that starts with some available parts—but novelty is different because of its consequences. Whereas a transformation can be the reduction of parts or a change in size or proportions, a novelty is a one-time evolutionary appearance of a new structure that enables the evolutionary success of all of its descendants. Today's whales are exemplars of the great success of evolutionary novelty, with echolocation in toothed whales and filter feeding in baleen whales enabled by biological apparatuses present in none of their forebears.

Evolution is the intellectual glue that connects living whales, in all of their seeming weirdness, to their deep ancestry, which is both incomplete and still not entirely known. Skulls point us to these evolutionary clues in a tangible and clear way; and without these insights, and the millions of years that brought these changes about, it would be very hard to illuminate the connection whales have with other closely related mammals.

My way back to whales, as a scientist, has always involved skulls in one way or another. In college I puzzled over a half-rotten dolphin head when our field class, on a barrier island off the coast of Georgia, discovered its carcass buried in the sand. The putrid smell dispersed the rest of the class, but I held fast. I wasn't riveted by the grotesque allure; instead I was captivated for the first time by the thought of how, exactly, a whale might become a fossil.

In graduate school, I sought fossil whale skulls peeking out of seaside cliffs or crumbling out of rock formations in badlands that were once seafloors. Cetaceans became my vehicles for understanding life over geologic time, across scales so vast that we cannot truly comprehend them, even if paleontologists casually discuss geologic markers as if referring to last week's dentist appointment. And it all led me to the Smithsonian, where I tend to the world's great collection of fossil whale skulls.

However, part of the deal with being a museum scientist, especially at the Smithsonian, is that people ask you for tours. It's more than a fair trade. First, I like giving tours. They give me an excuse to try out new ways to talk about ideas—the big ones, like the evidence for evolution and extinction—that not only excite me but also explain how whales came to be. Also, tours often involve children, and children are usually the toughest audiences of all, whether they're whale huggers or fossil fiends. If I can figure out how to keep them interested, even for a short show-and-tell, then I think I've done my job. And who knows, I might even be convincing enough to ignite more than a passing interest.

My friend and colleague Megan McKenna first taught me about whale heads and their inner biosonar anatomy when we were both in graduate school, so when she brought her family recently for an early-morning tour before the museum opened, I was excited at the opportunity to settle a debt. Her four-year-old daughter was just a bit older than mine. The museum's halls, especially when they're empty, can be intimidating spaces, so I started slowly—I didn't want to overwhelm or manufacture too much excitement.

As we walked into the Sant Ocean Hall, we stopped short underneath the imposing right whale model, and turned to face the eel-like skeleton of *Basilosaurus*, grimacing from above.

Basilosaurus is an early whale several million years younger than *Pakicetus* and *Maiacetus* but it looks worlds apart in size and shape. Its dinosaur-sounding name literally translates from Latin as "king lizard," in a nod to its serpentine, bus-length body. The first fossils of *Basilosaurus* were collected from the chalky marls of rural Arkansas and Alabama in the early nineteenth century. With a skull over three feet long, jaws bearing palm-sized, saw-shaped teeth, and individual vertebrae large enough to serve as stools, a *Basilosaurus* skeleton gives every impression of belonging to a sea monster—and its name has stuck, if only for the conventions that scientists use to name species.

Unlike today's whales, *Basilosaurus* has a head that does not dominate its body, arms that crook at the elbows, and a surprising, diminutive set of legs that could never have supported its large weight on land. But *Basilosaurus* has an involucrum like all the other early land-dwelling whales, and its ear bones floated in a sinus space below its skull. *Basilosaurus* did not echolocate, nor filter feed, leaving it caught somewhat in an evolutionary middle ground for whale evolution: one of the first fully aquatic

A Basilosaurus *tooth*

whales, completely unreliant on land, yet still carrying many of the biological apparatuses of its terrestrial ancestors.

As I walked with Megan under the skeleton, I pointed to its hind limbs and joked aloud about how they dangled like poorly placed, miniature landing gear. Megan then leaned down to her daughter. "Hey, Etta, Nick studies whale bones just like those ones." Etta looked at me intently, mulling the assertion. "Why?"

I opened my mouth to deliver a canned response involving school, science, and curiosity, more boilerplate than authentic. I knew I could do better. I waited a few beats.

"Their bones all tell stories," I said, "about where whales came from." She glanced up at *Basilosaurus* looming from the ceiling, like a giant, flippered, macabre snake. "And if you become a scientist, you can learn to read them and know their stories." I knew I had her attention. "But they sure don't just show up here in the museum all put together," I smiled. "You have to find them first."

3.

THE STORIES BONES TELL

I walked with my eyes trained on a long, gray road cut through a hillside on a cattle ranch. In the late-afternoon California sun, the waving golden grass gave the hill the look of an enormous, shaggy animal, revealing its sedimentary flank. I followed a small path, parallel to the exposed fossil-rich layer. Walk, stare, walk some more, scrape the exposure, and then walk again; maybe you get lucky.

A few yards away from me, my colleague Jim Parham was doing the exact same thing. I've known Jim since graduate school, and we don't need to say much to each other in the field. Jim's an expert on turtles and other reptiles. As it is for me with whale bones, Jim has seen enough specimens that even the smallest fragments of fossil shell can help him solve riddles about turtle origins, which stretch back even deeper in geologic time than whales, though we tend to find fossil sea turtles in the same type of rock as fossil whales. Jim and I reliably fall on the same page, by temperament and by rock units.

We had visited many other outcrops in the foothills of the Sierra Nevada together. We worked side by side, scanning in silence. "Hey," Jim said abruptly, reaching down. He raised a palm-sized shark tooth to the sky, its serrated edges cutting the orange light. I looked down and immediately started to see other shark teeth and whale bone fragments recently eroded out of the hillside, gems in the rough. "Oh, check this out," I said, retrieving a segment of dolphin rib from the newly formed sediment piles. As I flipped it over between my fingers, I noticed something unusual—a set of a dozen parallel lines gouging a path across the bone's surface. A shark bite. This site was, after all, the Sharktooth Hill bonebed.

The fact that the rib bone belonged to a small species of extinct toothed whale (Odontoceti indeterminate, if you want to be technical) was probably the least interesting thing about it. That kind of identification is merely born out of the same patient study—hours with museum collections—that gives you eyes for spotting bones in the first place. Far more interesting was the fact that it told us part of a story: a little more than fifteen million years ago, in the middle of the Miocene, an ancient shark chomped down on an extinct dolphin's rib cage.

Whether this particular set of bones represented a fatal encounter or mere scavenging on a carcass we couldn't know. There was also no real way of knowing whether the shark tooth in Jim's hand and the marked rib in mine were causally related. We held the two side by side, checking the serrations on the tooth with the gouges on the rib—close, but not a precise match. Even if they were, it would be a stretch to tie the two pieces of evidence, a gumshoe's leap in causality at a suspected murder scene. Whale bones do tell us stories, but they're not always satisfying or predictable.

I lodge finds like these on the shelves of whale bones that I keep in my head. I can't quite tell you how this mental library is organized, but I slip into it every time I see a shard or glint of whale bone, whether out in the field or in a museum drawer. The more fragmentary, the more fun. I pick it up carefully, thumb its creases, divots, and twists, and then scrutinize its topography by eye. My thoughts immediately race through a chain of mental flash cards to arrive at the best possible identification of its former owner: Right or left side? Symmetrical, from the main axis of the skeleton? Cranial or something below the neck? Scavenging marks? Pathologies? These flash cards are marked with names for every bump and hole on a bone's surface. It has taken me years to build up this cerebral collection, long hours spent with many skeletons within arm's reach, flipping each piece over and over again, tracing each surface for memory. It's also good to keep a stack of real literature on hand as a guide, because you certainly aren't the first one to pick up a whale bone and ask it a question: How did you get here? Where in the skeleton do you belong? What happened to your owner? There is an undeniable thrill in this chase, whether it's in the field or in a museum collection, and fortunately you carry your mental library everywhere you go. Anyone can participate too—amateur sleuths sometimes crack cold cases.

There is, however, a catch: you hardly ever get all the answers. As with other vertebrates, the fossilized skeletons of whales tend to be massively incomplete because the organismal glue that keeps skeletons together—stuff like ligaments, fibers, cartilage, and muscle—decays rapidly and is dispersed by waves, scavengers, and time. Our knowledge of most fossil whale species is based on little more than a battered skull, lacking all but the most diagnostic and unique features. For some time periods, and in some parts of the world, we can put all of what we know about the whale fossil record on one table. These fragments of bones—skulls, teeth, vertebrae, limb bones—can look like a jumbled puzzle waiting for someone to bring the missing pieces. Or, preferably, the cover of the box.

This situation is what we tend to find for most fossil whales from the first phase of their evolutionary history, the part that took place at least partially on land. We don't have complete skeletons for *Pakicetus*, *Ambulocetus*, *Remingtonocetus*, or most of the close relatives of *Maiacetus*. Being aquatic clearly helps in getting preserved intact. Perhaps the size increase from *Maiacetus* to much larger early whales such as *Basilosaurus* is part of why we tend to find more complete skeletons belonging to fully aquatic whales (while the bones individually get larger, they are fewer in number when you reduce and eliminate paired leg and foot bones). The fact is we don't have a good understanding of the intermediary steps between hind-limb-propelled whales to tail-propelled ones. For all the anatomical transformations that happened in the earliest whales, there is a gap in the fossil record and our understanding between the last semiaquatic and the first fully aquatic whales. To fill out the picture we need more fieldwork in the right places with rocks of the right age, and a lot of luck.

A good paleontologist can go far on scraps alone, and sometimes we're lucky. There are places—or times, because paleontologists think in both space and time—where the fossil record yields parts of hundreds and even thousands of individuals. These fossil-rich areas are called bonebeds. My mental library comes in handy when I encounter one, helping me distinguish a scrap as the bone of a whale versus that of any other animal. At their densest, fossil whales in bonebeds get jumbled with scraps of other extinct marine mammals, seabirds, sea turtles, and sharks into layers only a few inches thick. At the other end of the spectrum, complete whale skeletons can be distributed over a broad area that can even reach square miles. The definition of a bonebed has mostly to do with the fact that skeletal parts are concentrated within a single layer of rock. What paleontologists and geologists want to know, once they've found a bonebed, is how much geologic time has compressed the evidence, which can represent as much as a million years or maybe just a day's worth of a flood.

In the 1920s one of my predecessors at the Smithsonian, Remington Kellogg, recognized that the Sharktooth Hill bonebed in the foothills of the Central Valley contained a richness of fossil whales, mostly identified on the basis of broken skulls and individual ear bones. The bones that form the outer, middle, and inner ears of whales are among the most heavily mineralized bones for any mammal. All the better for hearing underwater—and for preservation in the geologic record. The acoustically isolated ear bones discussed previously in bottlenose dolphins can also be found back to the time of the Sharktooth Hill bonebed and beyond, to the age of *Pakicetus.*

Kellogg described and named twelve previously unknown fossil whale species from the Sharktooth Hill bonebed, encompassing a range of extinct baleen whales, early sperm whales, oceanic dolphins, and distant relatives of river dolphins. At this point in whale evolution, the world was full of filter-feeding and echolocating whales—all land-dwelling ones were long extinct—but they lived alongside sea cows, strange, hippo-like herbivores called desmostylians, early seals, and early walruses.

The material record of that past world comes from the Sharktooth Hill bonebed, which is an orange and brown layer only a few inches thick, chock-full of bone bits spread over a dozen or so square miles northeast of Bakersfield. Jim first introduced me to the bonebed in my early years of graduate school—he was more interested in its fossil sea turtles. Eventually I focused on figuring out the precise age of the bonebed and how this kind of dense rock unit, full of bone nuggets and occasional skeletal parts, came to be. Context is everything, and without it, answers to the bigger ecological questions about the past are undecipherable.

The bonebed was essentially an exposed seafloor for several hundred thousand years, collecting the hard-part remnants of Miocene whales, sea turtles, sharks, and other animals that fell to the seafloor while lighter sediment swept past. Those few inches of bonebed today thus capture a condensed interval of time, between sixteen and fifteen million years ago—not much for a geologist but a span much longer than the duration of our own species. It's also a span of time probably long enough to sample the full range of extinct whales and other backboned animals that lived in the vicinity of this part of California when the Central Valley was an embayment open to the Pacific. Knowing how many fossil whale species were around, giving them all scientific names, and understanding their evolutionary relationships is all ongoing work because there's so little skeletal material to use as a basis for a species. (Kellogg used ear bones for most of his species, a puzzling move given how limited they are for species-specific identification.) Such work is time-consuming and exacting, measuring and comparing scraps of bone to one another. Many times we're simply left saying, "This is something new, and it deserves a name, but we can't say more until someone finds a good skull."

Kellogg wrapped up his Sharktooth Hill work after finishing his dissertation and secured an appointment at the Smithsonian in Washington, D.C., where he turned his attention to fossil whales that were more complete, from an earlier time in whale evolutionary history. By the 1930s, the Smithsonian possessed the world's most extensive collection of early whales, but they weren't the land-dwelling ones, such as *Pakicetus*, whose skeleton still wouldn't be found for another half century. Instead, there were drawers and drawers of *Basilosaurus* and other species of the first of the fully aquatic whales, more than enough to mount full skeletons for exhibit halls, and sufficient to know something about what these extinct whales were like.

Basilosaurus hardly seems like a whale—saying it's almost like a whale would be charitable. It had a toothy, snout-dominated head, looking something like a gigantic leopard seal, except its nostrils were located not at the tip of its snout but about halfway farther back. It had a visible neck, unlike most of today's whales. While its fingers and hands were probably encased in flesh, forming a paddle, it could bend its arms at the elbow, as no living whale can. The most remarkable thing about it was its long, eel-like body—most of its length came from its tail. *Basilosaurus* probably had a tail fluke, but it also had cartoonishly small hind limbs. These hind limbs were vestiges from its land-dwelling predecessors; as mentioned previously, they could not have held up *Basilosaurus*'s enormous weight (about six tons) on land. In other words, *Basilosaurus* was fully aquatic, living its entire life underwater.

Kellogg knew only a little about those tiny hind limbs—the collections at the Smithsonian, for all their depth, still comprised only a pelvis and a single femur. Did *Basilosaurus* have feet? Toes? The answers to those questions eventually came from *Basilosaurus* skeletons found in Egypt, a world away from the coastal plain of the United States, and many years after Kellogg died.

Since the nineteenth century, paleontologists have known that the same strata used to build the pyramids of ancient Egypt harbor marine fossils, including fossil whales. These fossil-rich rocks crop out for nearly a hundred miles to the southwest of Cairo, exposed at their grandest scale in a place called Wadi

Al-Hitan, loosely translated as "Valley of the Whales," in the Fayum depression. Toward the end of the twentieth century, more detailed work in the area produced a species list including ten different early whales, along with early sea cows, primates, and the earliest elephant relatives. Wadi Al-Hitan, however, earned its name because of the fossil whales—especially *Basilosaurus*—whose skeletons number in the hundreds, spread across miles of desert expanse edged by cliffs and wind-battered mesas. These skeletons come from bonebeds, just in a different mode from Sharktooth Hill. Over three hundred skeletons of early whales are littered across one hundred square miles, including the first complete *Basilosaurus* skeletons ever found, with skulls, arms, rib cages and tail vertebrae, and legs—everything down to the four tiny toes intact. What these legs might have done in life, beyond being mere vestiges, remains unclear. (Some scientists have speculated that these small legs might have been used for copulation, especially given the animal's extreme snakiness.)

Complete skeletons of *Basilosaurus* give us plenty of clues about its behavior. Like some of its predecessors, *Basilosaurus* had acoustically isolated inner ears, letting it hear directionally underwater, but it lacked the anatomical space to house any kind of echolocation organ on its face. (*Basilosaurus* therefore heard only low-frequency sounds—not the ultrasonic ones that echolocating toothed whales use today.) It ate fish, based on fossilized stomach contents. Every tooth in the head of *Basilosaurus* was capable of crushing bone, and its overall bite force exceeded that of any other mammal, living or extinct, including hyenas. Bite marks on the skulls of another, smaller species of early whale from the Fayum suggest that *Basilosaurus* ate other whales, the way killer whales do today. One major difference with killer whales: *Basilosaurus* could crush its food, whereas killer whales rip and tear, oftentimes working together. At the moment, we can't say whether *Basilosaurus* moved about in pods—there's no good fossil correlate for that kind of behavior.

Basilosaurus

The fossil-rich rocks of Wadi Al-Hitan reflect ancient shorelines formed during episodes of periodic sea-level rise and fall at the end of the Eocene, around 40 million to 35 million years ago. *Basilosaurus* probably inhabited these lagoonal environments (it certainly was buried in them); it lived not unlike many dolphins do today, ranging from coastal shores to open water. By the time

Basilosaurus went extinct, at the end of the Eocene, subsequent branches in the whale family tree leading to today's whales had already evolved. While we don't have a good fossil record for the very beginning of today's echolocating and filter-feeding whales, we suspect they looked a fair bit like *Basilosaurus*—fully aquatic, though less snaky, sized-down shadows of the leviathans that they would later become, tens of millions of years later.

So much for the hows of fossil whales with legs. But why? What led whales to return to the water from land in the first place? That question takes us to the gap between the first and second phases of whale evolution, the gap that remains in the family tree between the branches leading to *Maiacetus* and *Basilosaurus*. In about ten million years, whales went from looking like the four-legged *Pakicetus* to something closer to *Basilosaurus*. Sometime during that interval (and probably in the last half of it), whales ambled and swam equal amounts, with shorter hind limbs and blowholes migrating backward along their snouts. And then, at some point, a generation of whales never emerged out of the water back onto land, and their descendants begat blue whales, humpbacks, sperm whales, dolphins, and every other living whale species (along with many extinct ones, like Kellogg's finds from the Miocene).

The search for true causes—especially in the evolutionary sciences—is usually not as conclusive as the search for patterns and their data. Hows are much more forthcoming than whys. For whale origins, multiple explanations for their reentrance abound: they returned to escape predators on land; to take advantage of more prey at sea; to seize new habitats unexploited by any major marine predator since the demise of gigantic marine reptiles at the end of the Cretaceous, about twenty million years prior. Each one of these explanations is plausible but difficult to test. Maybe we'll one day refine those explanations into a hypothesis with a prediction that we can evaluate, perhaps using the geologic context of these early whales, comparisons of their osteology with those of marine reptiles, or a novel analytical tool. One thing for sure: we will certainly benefit from more fossils—so we should keep looking.

Every scrap of fossilized bone found in the field may be novel, but they're not all precious. There's always some decision making about whether any particular fossil should be collected in the first place. It is, really, all about the questions at hand and how any certain fossil find can help answer it. Bonebeds are like caches of evidence: areas rich with clues, either because of the density of remains contained within them, as at Sharktooth Hill, or because of the completeness of the specimens in a given space, as at Wadi Al-Hitan. One fossil find can tell us about that individual, but it also captured a snapshot of a real ecological interaction, lost in geologic time. That's an important detail from life in the distant geologic past, especially when we want to know the details of not merely the anatomy or evolutionary relationships of extinct organisms, but the food webs and ecosystems in which they lived millions of years ago. Finding these kinds of paleontological caches is thrilling, and it can also be overwhelming, as I would soon find out for myself.

4.

TIME TRAVEL ON THE
FOSSIL WHALE HIGHWAY

Imagine floating above the great tapered tail of South America from space, seeing it stripped of clouds, ice, soil, and water so that the geologic world beneath is made visible. The familiar outline of the continent rises in stark, jagged relief. The high spine of the Andes is draped in red and gray bands to the east, toward Argentina, and ochers and sands cover Chile to the west. From this vantage, the cone of South America is locked in by a jigsaw puzzle of oceanic plates, and a surprisingly deep, dark cut mars its western boundary.

This incision marks the border between the Nazca and South American tectonic plates, where the lip of the former inexorably and slowly rolls under the edge of the latter. This action uplifts what was once seafloor, carrying ancient organisms buried in it—extinct whales included—slowly to dry land on the western edge of South America. This tectonic motion, called subduction, eventually yields mountain chains like the Andes over geologic time. But at the scale of human lifetimes, subduction can cause megathrust earthquakes that convulse entire cities, maroon fishing boats, and kill thousands in a span of seconds.

In 1835 a young Charles Darwin observed the outcome of this

very process along the coast of Chile, near Concepción. Three years into its round-the-world voyage, the HMS *Beagle* had rounded the horn of Tierra del Fuego and made its way up the west coast of South America. Darwin was ashore when the earthquake started. A crescendo over the course of hours allowed most of the residents of Concepción to flee and limited the scope of fatalities to a few dozen people. Aftershocks rattled terrified locals for several days thereafter. Darwin later surveyed the devastation in Concepción firsthand, noting that most of the city was flattened, burned, or flooded by an accompanying tsunami—and that the entire shoreline of the harbor had risen several feet, stranding limpets and starfish. Darwin surmised that these catastrophic effects were connected with the volcanic eruptions he had observed during earlier forays hundreds of miles south near Chiloé. Darwin suspected that volcanic eruptions, the sudden uplift of coastlines, earthquakes, and tsunamis were linked by a common underlying mechanism. His intuitive guess was more right than he could have known; they are all consequences of subduction—the jerky slippage of great masses of tectonic plates against one another—and the central process that underpins the idea of plate tectonics.

Plate tectonics is a very young idea about how the Earth works. Until the late twentieth century, geology textbooks did not have a clear answer for why South America's eastern edge fit so nicely with the west coast of Africa, which is a bit like launching moon-bound rockets without knowing Newton's physics. Eventually scientists discovered that convection currents from deep inside the Earth drive the fragmented crust of the Earth's rocky surface into constant motion, over geologic time. Every continent, and the ocean plates between them, floats on a vast, molten, and churning globe. The idea of plate tectonics also neatly explains a variety of patterns in the fossil record, including why so many plants and extinct animals across the southern continents look so similar—namely because they were once, a hundred million years ago, living together on a larger continental mass that has since broken apart.

Before he started thinking about evolution, Darwin was a geologist, one with a long view of history and the planet. Deep Time was a new idea when Darwin roamed the South American cone. What he saw there—the earthquake at Concepción, petrified forests in the high Andes, and fossils of extinct land mammals in Patagonia—resonated with the concept of an unfathomably old planet, one that had actually weathered many billions of years, time enough for the power of selection to yield finches, tortoises, and whales.

Darwin spent his last days in South America on horseback in the Atacama Desert of Chile, geologizing away at any exposure of inner Earth while en route to meeting the *Beagle*, which was moored along the northern coast near the town of Caldera. From there he headed north and westward by sea, eventually to the Galápagos Islands. Textbooks celebrate Darwin's few weeks on the Galápagos but tend to minimize the fact that he spent two years in the South American cone. He never returned to Chile, but his social network continued the work he had begun. Correspondences born out of a shared love for the land, and scientific questions about it, roused a generation of Europeans who decamped for its rugged, open landscapes. The result is a testament to the strength of friendships that can transcend generations: they established the first centers of learning in Chile, including a national museum, whose collections still hold fossils that Darwin collected over 180 years ago. When you open up museum drawers and handle those specimens, you realize that these physical objects connect scientists across centuries, anchoring the questions that we have about evolution on this planet.

Sweeping offshore of the Atacama is the Humboldt Current, an enormous, ever-flowing body of water that cannot be discerned by the naked eye. Its namesake is a scientific figure of impressive breadth and accomplishment who preceded Darwin by decades— although Alexander von Humboldt never made it south of Lima, Peru. Today the Humboldt Current is renowned as one of the richest fisheries on the planet. Open any can of anchovies or sardines, and there's a good chance that its contents came from the ocean current that stretches from Chilean Patagonia to Peru and the Galápagos.

Understanding how the Humboldt Current works, in the simplest terms, requires a step back to look at the coastal phenomenon of upwelling. When the Earth spins on its tilted axis, hot air rolls unevenly off the continents and transforms into trade winds over open ocean. These winds push hot water away from coastlines, and in its place arise enormous spirals of nutrient-rich water from the ocean's depths, in a process called upwelling. Westerly coastlines across the world, from California to Chile to Angola, all share the right set of geographic features for this process. Upwelling forms the foundation for rich and productive ocean food webs because of the nutrients brought up from the deep: ocean water at the surface generally has enough oxygen, but it's upwelling that carries nitrogen and phosphorus into shallower waters and enriches them by fertilizing both light-fixing phytoplankton and their consumers, zooplankton. The oceans may be vast, but upwelling creates specific places where these tiny organisms gather. Where there's food to eat, that's where whales, sardines, and penguins all want to be.

And if you're a whale paleontologist, you want to be on the coast nearby, where upwelling and subduction combine in a perfect way. Upwelling gives us whales, and ultimately their remains—whale bones—to look for, while subduction uplifts ancient seafloors to dry land. Along with these two processes, the accident of the Atacama's latitude makes the entire desert a gift for scientists seeking rocks: it's devoid of grass, trees, and the blacktop of civilization. Arid and exposed badlands allow erosion to exhume the remains of ancient whales locked in rock without the interference of soil or tree roots.

Erosion helps, but paleontologists also have to find the right kind of rocks. Of the three categories, igneous, metamorphic, and sedimentary, for fossil whales only sedimentary rocks will do. Whales don't preserve well in volcanic lava flows, and fossils hardly ever survive the tremendous heat and pressure that creates metamorphic rocks dozens of miles beneath the Earth's surface. Among the array of sedimentary rocks, the most promising ones include mudstones, representing offshore seafloors, and sandstones, representing nearshore ones. Thanks to uplift, the only way to find fossils in the Atacama involves rolling across washes and mesas by foot or truck, finding these rocks, and keeping an eye out for a glint of bone. Whale bones tend to be relatively big, after all.

Since Darwin's time, many thousands of bits of bone and teeth have been collected from the Caldera Basin, some ending up in natural history museums such as the one in Santiago, founded by Darwin's correspondents. Fossil collections from Caldera consist almost entirely of fragments of skulls, limb bones, and teeth—never complete skeletons. But the little they show suggests that the Humboldt Current of the past was different: familiar species such as whales, sharks, and sea turtles lived alongside bizarre extinct ones, including nightmarish bony-toothed seabirds, long-snouted aquatic sloths, and school bus–size predatory sharks. When I started to contemplate working in the Atacama, the problem was that we didn't have precise ages on all of these fossils in hand—we needed one complete chronicle of rocks in the Caldera Basin. Minimally we needed to identify the correct succession of layers these fossils came from, identifying oldest to youngest; at best, we hoped to pin down the ages of each fossil-bearing layer with numerical geologic dates. Once that context was arranged, we would be able to chart the rise and fall of each species over millions of years, against the backdrop of broader changes to ocean temperature, sea level, and circulation. Wear a geologist's hat, find whale bones, and figure out the story of how the Humboldt Current's ecosystem came to be.

After years of planning, correspondence, and false starts, I found myself sweating in the glare of the Atacama sun. I paused from scanning a geologic map of the Caldera Basin, spread across the hood of the Toyota pickup, and squinted, hoping to catch a glimpse of people summiting the flat top of a mesa in the distance. The harshness of the white light pinched my vision, split between the patchwork of colors on the map before me and the pale-blue dome above my head. I was frustrated, my mind elsewhere. We were late. The students on the field team hadn't reconvened at the agreed-to time and we needed to keep moving.

We had followed the whale bone trail in the Atacama, and the trail led us straight into fault lines—many of them. Finding fossils wasn't exactly our problem. Instead, it was their context. We were having a difficult time piecing together the succession of rock layers in which we were finding them. As tectonic processes uplifted ancient seafloors, they also broke them up, like a layer cake dropped on the floor. Consequently, deciphering the specific old-to-young sequence of layer upon layer of rock was complicated by long, vertical fractures that displaced the layers up and down relative to one another. We tend to think of geologic faults as stretching across hundreds of miles, and that's certainly true for some of them. But they also manifest locally, in a rock outcrop that may be no broader than the side of a house. In the Caldera Basin, faulting sometimes created a jumbled layer of rock rather than a neat stack.

With boots on the ground, building a single chronology of rocks meant finding specific places where we could measure the thickness of the rock layer in a repeatable way, using a simple geologic tool called a Jacob's staff. We would also note the composition, color, and texture of each rock layer. Occasionally we would also hammer out a sample of the most promising rocks— usually the ashes—in the hope of finding tiny volcanic grains that would yield precise geologic dates back in a laboratory. Through this slow, exacting work of measuring, describing, and sampling, we hoped to pin down actual dates in geologic time for enough layers in the sequence to understand the succession of different extinct species, whales and otherwise, that once lived in the Humboldt Current.

Back at the truck, however, I wasn't thinking about geologic maps or envisioning layers of fossil whales captured through time. Instead, I was thinking about hours wasted, miles away from the air-conditioned convenience and sprawling desk work within my museum's walls. I thought about all of the effort and time—coordinating airfares and truck rentals, pushing permits along, accounting for family and professional commitments. As the students crested the mesa, I waved at them. What I really wanted to do was slam on the truck's horn until it was out of air.

Carolina Gutstein, my friend and colleague, then finishing her doctoral degree, stood with me at the truck. "You know, you can't just rush people," she leveled, without hesitation, sizing up my agitation like a sibling. "Trying to make things go faster here is only going to make them go a lot slower." I laughed ever so slightly but stopped when I turned to look at her. Caro's face was dispassionately still, her mountaineering sunglasses reflecting a stereo double of my own weary glance. I looked down in frustration, back toward the map on the hood, and exhaled. When I glanced her way again, she smiled, breaking the tension. "Why don't we go see all those whales at Cerro Ballena?" she offered. "I'll call Tuareg to show us around. He and Jim are over there now. You're not going to believe it."

Actually, I thought that I had good reason not to believe it, especially if it involved the man who calls himself Tuareg. His real name is Mario Suárez, and he is probably the single best fossil finder I have ever known. He demurs when asked, but Mario's self-appointed nickname is clearly meant to evoke the stoic Berber people of the Sahara—an image he routinely betrays by losing his cell phone (he has lost more than a dozen) and completely going AWOL when needed (usually found at the nearest bakery). But at the time, we were strictly in his domain, working under his collecting permit, which he held as curator at the local paleontology museum in the town of Caldera.

Tuareg had e-mailed me earlier that year about a place he started to call Cerro Ballena, where he said complete whale skeletons had been found, but I'd had a hard time discerning much from afar. I remembered having seen the site on a past visit, a sloping road cut of the Pan-American Highway that trenched through a layer cake of orange and tan marine rocks. The only fossils I had noted were a smattering of skull bones from a large whale, likely a baleen one. Locals had tried tunneling out the bones, unsuccessfully, next to graffiti carved in the soft sandstone. None too auspicious. Fossil whale skulls are sometimes a jumble of broken bones that hardly make sense at the rock outcrop and require care and study back at a laboratory. Also, they almost always involve heavy logistics that simply outstripped our time, our resources, and, to be frank, my motivation.

Caro's suggestion reminded me of that whale skull we had seen together by the side of the road, although I was only burdened by the recollection to a point. If we had more time, maybe we could collect it, but we had to make hard decisions on the use of our time. We were in the Atacama to understand the evolution of the Humboldt Current ecosystem, as read from layers of fossils from dozens of species across time—doing so offered the chance to find out much more than a single broken-up baleen whale skull could ever tell us. Constructing a single stratigraphic column from the stacks of rocks across miles of fractured desert terrain was something reasonable and feasible to achieve during a single field season, if not particularly sexy. I was also on the hook to deliver publications out of our work, as a foundation for future collaborations. As it turned out, I had no clue how wrong I was about the importance of that broken-up skull at the side of the road, nor a hint about the scope of what it represented.

If I was ambivalent about seeing Tuareg, at least I was buoyed by the thought of reuniting with Jim Parham. Jim is a like-minded scientist, a friend, and a terrific sounding board. His finely tuned bullshit detector always checked my field decisions about logistics. Earlier that day, we had split the field team in two to maximize our time: Caro and I took the students to the south, while Jim and Tuareg went north to Cerro Ballena. "I really don't think we *both* should be in the same truck as Tuareg," I said to Jim at breakfast. "Oh—just as a matter of sanity," he assented.

When Jim, Caro, and I had visited Cerro Ballena with Tuareg several years earlier, we'd referred to its location as "that road cut next to Playa del Pulpos," taken from the nearest highway sign. By late 2010, it had become Cerro Ballena—literally "whale hill" in Spanish—if only because of global geopolitics, in this dusty part of the Atacama. In the past few decades, Chile's geologic resources have become prized targets for extraction by the mining industry, and accommodating large mining machines has meant road widening along remote parts of the Pan-American Highway. An environmental-impact study at Cerro Ballena concluded that further expansion would very likely uncover more fossils. Nevertheless, a road-construction company was greenlighted to begin widening the highway. To comply, the company enlisted Tuareg and his museum for assistance, with Chile's strong natural patrimony laws ensuring that any fossils would be saved. It was then that Tuareg had started sending me clipped e-mails and shaky videos from the site, not exactly adequately conveying the message of what was happening there. Besides, it was Tuareg—hard to pull out the facts from hyperbole.

When Caro and I arrived, Tuareg and Jim were pacing about the quarry. Large black felt tarps dotted the desert floor every dozen feet, stretching north and south. I ambled up to Jim. "Dude," he said in a low voice, telling me everything I needed to know in a single word. "This is not the Playa del Pulpos that we saw two years ago." Everyone gathered to follow as Tuareg walked from tarp to tarp, rolling each one backward. My mouth fell open as I absorbed the fact that every tarp covered at least one complete whale skeleton, and sometimes several on top of one another. Every black tarp, dozens spread up and down the road-cut quarry, demarcated a whale skeleton. The sheer density of complete skeletons outstripped everything I thought that I knew about how whales get preserved as fossils.

The skeletons, some thirty feet long, were almost all complete in a way that fossil whales hardly ever are, nose to tail. Many looked as if the creature had died in place, carefully turned on its back, and then been pressed flat over geologic time, like a preserved flower. Skulls were easy to spot, their triangular projections and bowed jawbones at the end of a trail of bricklike vertebrae. Rib cages collapsed toward tails, like gigantic Slinkys. In many skeletons, the ribs were still adorned with shoulder blades connected to arms and even finger bones. The fossil whales at this site were jaw-droppingly complete. And it made no sense that there were so many, so close together. I couldn't think of any other field site of fossil whales like it.

I was stunned. Tuareg gabbed away with a positively gleeful Caro and her students. I walked over to where Jim stood at the south end of the quarry, taking photos and rubbing sediment between his fingers. We silently watched the sun slipping over the horizon, evening cloud banks bringing a cool wind. In the distance, a single round peak—El Morro, a weathered mound of igneous rock—capped the view.

"It's over," Jim said, flatly. I looked north and south across the entire quarry, more than a football field in length. I knew exactly what he meant: anything we had thought about doing needed to make room for this site and the several dozen skeletons that stretched up and down the hill in each direction. Measuring the stratigraphic columns across the Caldera Basin, slotting in all of the fossils we already knew about, deciphering that geologic map full of faults—it all needed to wait. My hours and hours of planning had focused on a sure thing, returning with bags of rock samples and with notebooks full of the makings of promised papers. Entire whale skeletons were not part of that plan, certainly not dozens of them.

I breathed in, anxious and unsure. I was frustrated with myself for not listening to Tuareg more carefully earlier on, paralyzed by the enormity of the scene in front of me. At the same time, part of me recognized that the scope of the site, with its dozens of perfect whale skeletons, was undeniably significant—and I had an open invitation to be one of the first ones to study a place like nowhere else, as far as I knew, on the planet. It was vexing and tantalizing; it was a kind of Pandora's box, and we'd just seen it crack open.

"I know," I said. "What are we going to do?"

5.

THE AFTERLIFE OF A WHALE

I think a lot about how whales die. That might sound like the ranting of a whale bone chaser gone full Ahab, but my preoccupation is not with the gore of decaying flesh or exploding body cavities (although those don't really bother me). Instead, I'm fascinated by the details of the what, where, how, and why: what happens to their carcasses, their locations when they expire, how whales perish, and the reasons for their demise. You might think that these facts are easily uncovered in the scientific literature, or in the many accounts of whaling on the high seas. But they aren't, not for all of the whales that have washed up on the world's shores or been hauled up by whalers. So I parse these factors in my head instead—ocean currents, water depth and temperature, scavengers, time to burial, and even anatomical differences—that contribute to the many different ways that a whale carcass might become a fossil.

Figuring out what parts of the living world can be entombed in rock, and how we might find them, is a game of probabilities. Paleontologists tend to think about the life and death of organisms as a continuous thread from birth to death—and to museum drawer. We visualize this thread as a stream of information where a variety of biological and physical processes winnow away data at each step in a pathway of decay: a carcass scavenged to pieces, not buried intact; a skeleton, or parts thereof, coming to rest in an unpromising setting; the rocks containing the fossil accidentally destroyed. Even if a great specimen is uncovered, the fossils may lie silently in a museum, collected from the field but mislabeled or undiscovered in a drawer. The reality is that we lose information throughout this process; it's an attrition of data from carcass to cabinet. Given the chances against any living thing becoming a fossil, it is a wonder that we know anything about life from the geologic past at all.

Thinking like a paleontologist makes you something of a connoisseur of dead things. My pursuit of dead whales has led to the rich record of strandings. Since antiquity, whale strandings have captivated the interest of everyone from Aristotle to casual spectators of exploding whale videos on YouTube. Strandings are a timeless motif—an immobilized, beach-cast leviathan, angrily tail-slapping against the surf. The image shocks us because we imagine whales to be fully a part of the aquatic realm. How would a whale end up landlocked in our world, a creature so large and strange suddenly so uniquely vulnerable?

Whale strandings happen in many different ways, for multiple reasons. Consequently, there is no single definition of a stranding—just an operational one, for the seemingly aberrant sight of a whale on a shoreline. For example, a stranding might consist of one whale, a mother-calf pair, several individuals from a single species, or several individuals from different species. Adding to the complexity is how they strand: whales may be dead upon arrival, alive but flailing on the shore, or already decayed to a raft of blubber, cartilage, and bones.

Beyond the how, there is still more complexity to the why: what causes a whale stranding? Senescence or disease may provide simple explanations in some cases, whereas the side effects of living near humans can be either plainly obvious (entanglement in fishing nets or ropes) or more difficult to plumb (toxin poisoning from marine algae). Certainly the sight of an entire pod of whales stranded, dozens in a row, begs some kind of explanation, although that is frequently elusive. True cause is often like that, in the natural world.

For naturalists working before the era of Yankee whaling in the midnineteenth century, strandings provided the only source of anatomical knowledge about whales. Despite all of the whales killed by Basque whalers for hundreds of years in Europe, there was essentially no documentation of what a whale looked like on the inside—fairly important evidence, when you consider that despite a few snout hairs, nipples, and nostrils for breathing air, whales otherwise largely look like fish on the outside. There are similarly limited firsthand descriptions of dissections on stranded whales from this era—though they must have been uniquely awful. Once word of a whale stranding arrived at the door of a rural doctor or an amateur naturalist in the eighteenth or early nineteenth century, the opportunity would have launched a whirlwind of ad hoc planning for several days of dreary, odorous dissection. The happenstance of a whale stranding dictated where the dissection occurred. The scale of the carcass would have dwarfed the tools at hand, the decay of flesh accelerated by nice weather or retarded by wet and cold conditions. The work was certainly not glamorous. And there were no modern winches or cranes, nor photographs to document the findings. Just ink, paper, and a strong stomach.

A stranded whale affords a detailed look not merely at diagnostic traits—a ridge along the snout, a piebald underbelly, or a knuckled tailstock—but at its inner anatomy, musculature, and organ systems, which can't be described from a boat. In the early nineteenth century, the first naturalists to roll up their sleeves and describe what they saw were enabled by an emerging infrastructure for scientific reporting—published scientific proceedings. By writing down, illustrating, and sharing what they saw, they created the basis for others to seek out their own comparisons. Even Aristotle knew that whales were mammals, but these first detailed dissections revealed facts of their inner world that were equally familiar and puzzling: they had a heart, lungs, a stomach, an intestine, and a reproductive tract just like a dairy cow or a tax collector. The early scholarship generated by careful anatomical work on whale strandings had serious impact on science, surpassed only by anatomists working under more stable laboratory conditions, with the benefit of refrigeration and power tools, generations later.

While today we know that, for instance, blue whales off Ireland, California, and South Africa all belong to a single species, naturalists in the eighteenth and nineteenth centuries did not. With incomplete (and sometimes incorrect) descriptions of other large whales, naturalists puzzled by variations in color or size would sometimes create a new scientific name based on a single stranding, or judge that a whale's appearance far from another record merited its description as a new species. It took until the early twentieth century for Frederick William True, one of my predecessors at the Smithsonian, to unravel these issues for large baleen whales and demonstrate that blue whales, humpbacks, and fin whales, among several others, were the same species on both sides of the Atlantic—despite dozens of taxonomic names purporting otherwise. True spent years working with original name-bearing specimens (called type specimens) for these different species, doing what taxonomists largely regard as housecleaning—a time-consuming task involving chasing specimens that are archived around the world's museums and figuring out their identity.

Even today there are some species of beaked whales known only from skulls washed up on a beach—yes, in the twenty-first century there are several ton-heavy species of mammals in our planet's oceans whose scientific basis primarily relies on a single beach-cast skull. Beaked whales are among the deepest-diving whales, looking something like a bottlenose dolphin crossed with a submarine. In fact, we know very little about most species of beaked whales, which account for nearly a quarter of all whale species alive today—they simply live too far at sea, dive too deep, and are incredibly difficult to tag or photograph in life. Without museums to house the rare remains that do turn up, we would know far less about these enigmatic species.

Not every whale that dies washes ashore, of course. Whalers for several hundred years have known that some whales float after death while others sink. Dead sperm whales float because of the enormous oil chambers housed above their faces, as Yankee whalers knew well. Right whales earned their moniker because they were the right whales to hunt, and they float after death because of their massive blubber layer, a trait they share with bowhead whales, their close relatives of the Arctic. Other large baleen whales, such as blue or humpback whales, will sink after a prolonged time at the surface, although carcasses can refloat following enough decay, when the gases from decomposition make the carcass buoyant. It isn't uncommon to see the large throat pouch of some of these whales balloon after death, like an emergency air bag that somehow failed to deploy properly in life.

Beyond these facts, known mostly to whalers and beach-combers, no one really knew much more until 1977, when a U.S. Navy submarine cruiser accidentally discovered a gray whale carcass on a seafloor more than four thousand feet deep, west of Catalina island, off the California coast. Of course, we already knew that some whale carcasses fall through the water column and reach the ocean floor, well beyond the depth that light penetrates—it's just that no one had ever seen the result until then. The scientists who later worked up the growing number of these discoveries called them whalefalls.

At thousands of feet deep, the seafloor is not merely cold and bathed in black; its surface is mostly barren—until a carcass lands, ending its transit through the water column, an elision between two worlds that began with the whale's last breath at the surface. Whatever flesh that has not already been picked away by sharks or pecked by seabirds provides immediate food for scavengers, such as deep-sea sharks, fishes, and crabs. (How, exactly, they find a fallen whale remains a mystery.) In very little time—researchers estimate weeks to months—these creatures will strip the carcass of its flesh, leaving only bone. On the deep seafloor, there's little current to disturb the position of the bones, leaving the skeleton looking mostly how it looked as it fell through the water column: the jaws close to or in direct articulation with the skull, which itself is connected to the vertebral column, in a straight line, with arm and flipper bones off to each side, assuming these parts weren't ripped off by scavengers at the water's surface.

But once seafloor scavengers swim and scurry away, the story isn't over. Scientists aboard deep-sea submersibles have set out in search of whalefall skeletons and even experimentally sunk whale carcasses to predetermined locations to learn more. With enough replications and time, they found that whalefalls undergo successive phases, not unlike a forest ecosystem that changes in composition and size as it matures over decades.

Once defleshed, whalefalls undergo a second phase of colonization by snails, clams, and polychaete worms—some feeding off cartilage and the surfaces of the bone, others burrowing into the apron of sediment around the skeleton, enriched by the organic material leaching off the whale's blubber and oil. The snails, clams, and polychaetes take months to a few years to consume all that they can, and afterward a third phase begins, which can last decades or more (no one knows because whalefalls have been studied for only forty years). This presumably final climax stage involves two sets of bacteria living in or on whale bones: anaerobic bacteria that use sulfate in the seawater to digest oil locked in the whale's bones; and then sulfur-loving bacteria that use the sulfide by-product of the anerobic bacteria to generate energy by combining it with dissolved oxygen.

Sulfophilic bacteria support a variety of true whalefall specialists at this stage, including some mussels, clams, and tube
worms that have the bacteria living symbiotically within them,
giving them the opportunity to generate their own energy in a
world devoid of sunlight. At these depths, whale carcasses give
a second life to an otherwise barren, abyssal world.

While the precise duration of these skeletons on the seafloor
remains unknown, the upper bounds of some estimates suggest
that a single whale carcass can provide up to one hundred years
of sustenance. So little is known about the breadth and variation in whalefalls that new discoveries are being made all the
time: one is an organism called *Osedax*—literally, Latin for
"bone devouring"—a species of deep-sea worm whose entire
life cycle depends on whalefall skeletons. Appearing as pinkish
filaments only a few millimeters long covering the surfaces of

bone, *Osedax* does not have a mouth or a gut, just wavy tendrils called palps facing outward. Instead of harboring symbiotic bacteria that use a sulfur-based pathway for decomposing bone lipids, its symbionts are a type of bacteria that mobilize proteins directly from the bone itself by dissolving it, using a tangled mat of bacteria-filled roots burrowed into the bones.

Not all whalefall colonizers are specialists; some are generalists that also make appearances on hot vents and methane cold seeps deep on the ocean floor. The range of the temperatures and environmental settings across these deep-sea habitats has led some scientists to argue that whalefalls, over millions of years, have served as evolutionary stepping-stones for invertebrates living in one habitat to leap to another. This idea remains hotly debated, with little known about all of the species that feast on fallen whales, or how often and where these skeletons are likely to appear on the seafloor.

A whale's size would seem to play an important role for a unique ecosystem that is fundamentally tied to its carcass. After all, a larger dead body should provide more opportunities for whalefall specialists. It turns out that size doesn't make too much of a difference where whalefalls are concerned, and the reason why we know has to do with the fossil record. As a graduate student, I had the good fortune to come across a fossil whale skull that had been collected a few decades prior, from rocks exposed on Año Nuevo Island, off the coast of central California. These rocks represent extremely deep-sea sediments about fifteen million to eleven million years old, and I didn't think much about the fossil's context until I was cleaning the skull in the fossil preparation lab at Berkeley's paleontology museum and came across tiny clamshells nestled in crevices of the skull. They clustered together, almost lifelike, and I decided to pry one off for a closer look, after first documenting their arrangement. A mollusk specialist confirmed a possibility that had entered my mind: they were chemosymbiotic clams belonging to a family that specialized in whalefalls. In short, the fossil I had been preparing belonged to a fossil whalefall.

Fossil whales with whalefall mollusks affixed to them had been found before and, while unusual, they weren't earthshaking discoveries on their own. But what was different was that this skull belonged to a whale that would have been barely eleven feet long in life. Tiny baleen whales—much smaller than those today—were common during the Miocene, but the remarkable aspect of this find was that their small body size did not prevent the whalefall from reaching the peak phase of sulfur-loving invertebrates. In other words, size doesn't really determine the community that colonizes a whalefall ecosystem. If not, then what does? That's still not clear, although it may be something about the lipids locked in the bones that controls which species can colonize the carcass, along with the stages of whalefall succession.

If you've ever seen a fossil on display in a museum, you might wonder why it is that animals sometimes preserve nearly intact, while others leave only a single bone. Understanding how the dead enter the fossil record is a field of study unto self, known as taphonomy, and it focuses on that entire pathway that filters the information we can know about an organism, from death to discovery. Taphonomy is really the study of information loss in anatomy and ecology. Ideally we want the whole picture of the ancient worlds that we study, but we don't ever truly get that, because of the vicissitudes of how living things fall apart after death.

Taphonomy has Old World origins, having been independently developed by Russian and German scientists working in isolation for the first half of the twentieth century. It wasn't until translations of their work trickled into the hands of American paleontologists, decades later, that the idea of using the biological present to understand death, destruction, and preservation in the past became a mature scientific field worthy of a name. One of taphonomy's pioneers was Wilhelm Schäfer, a German who spent decades observing the patterns of death and decay along the shores of the North Sea. While Schäfer patiently gave every sea creature that appeared as flotsam the same kind of attention he reserved for stranded whales, he nonetheless led off his seminal treatise on taphonomy with a decaying porpoise. Exacting and precise, he recognized the value of watching decay and decomposition patterns in large organisms, showing, for example, how a jaw peels off from a skull before the skull unlocks from the rest of the carcass. Whales are put together just like most backboned animals, and true to form, their lower jaws sometimes scatter far from the original carcass. This kind of observation is exactly the kind of clue that helps someone like me imagine how dead whales end up arranged as fossils—and how those fossils looked when they were whales.

Stranded whales always seemed to me like a fruitful place to start, but it took me some time to realize that you needed to think at the scale of oceans to understand what was important about them. In graduate school, a colleague of mine pointed out to me that nearly every species of whale that lived along the California coast had, at one time or another, washed ashore along a single ten-mile strip of coastline at Point Reyes National Seashore. When I went to dig deeper into his assertion, I found records kept by marine mammal stranding networks, coordinated by government agencies with federal oversight, which compiled whale stranding statistics for the entire West Coast of the United States, among other regions. Species identification, length, sex, condition—all data that were logged in the spreadsheets—provided an inventory of which whales (and how many) were stranded across nearly 1,300 miles of coastline. Interestingly, whale biologists working for these same government agencies had performed detailed survey transects tabulating whale species from boats, which led me to wonder how well these two kinds of observations—one from the dead, the other from the living—matched up.

The answer: surprisingly well. The dead and living data sets mirrored each other in terms of the number of whale species and their relative abundance—that is to say, the high proportion of individuals in some species over others. (For a variety of reasons, there are a lot more bottlenose dolphins than blue whales out along the coast.) In fact, over the course of decades, the stranding record recovered more species of whales than any living survey, including both common and rare species. In some cases, strandings picked up species that had never been seen in any boat-based survey. In other words, real ecological data are recorded by whale strandings, so long as you look at the right scales of time and space.

All of these thoughts spun in my mind as I walked through the rows of whale skeletons by the side of the highway at Cerro Ballena. I could imagine the site representing some kind of stranding. In the same breath, however, I also wondered if it was too tempting a label to apply, given that we didn't have any hard data in hand, nor any putative cause for why a stranding would have occurred here.

Strandings are rare in the fossil record; in the published literature, there is perhaps one possible mass stranding implied by piles of fossil ambergris (a hardened mass composed of squid beaks) preserved closely together, and another implied by three sperm whale skulls found together in a sand berm. Neither of these compared with the scale of what Cerro Ballena seemed to represent. Moreover, shorelines tend to be energetic environments where waves would disperse and destroy stranded carcasses—by contrast, many of the fossil skeletons at Cerro Ballena seemed undisturbed, hardly ravaged by the elements or by scavengers. From a brief walkabout at the site, I thought that Cerro Ballena had many outward signs of something like a mass stranding—especially in terms of the completeness and density of whale skeletons. How we could know, and sort out any possible explanation, remained a question very much at the top of my list. We were going to need data, lots of them.

6.

ROCK PICKS AND LASERS

In nearly twenty years of digging whales on every continent, the number of times I had ever found a complete skeleton of a fossil whale was exactly zero. It's not to say that it doesn't happen for other people—but it certainly isn't routine. The rarity of a single, complete fossil whale skeleton is a major aspect of why Cerro Ballena was so confounding: any one of its beautifully preserved skeletons would have been a discovery worthy of celebrating (and agonizing over its salvage); dozens of them yards apart demanded some kind of explanation. And a plan.

Up until I realized the full scope of Cerro Ballena, my fieldwork in Chile had consisted of several years of low-risk science focusing on paleontological essentials: measuring the thickness of fossil-bearing rock sequences at different places throughout the heavily fragmented Caldera Basin and connecting these blocks by linking rock units of similar type and appearance. The end goal was understanding how the Humboldt Current was different in the past.

Playing it safe was a sure bet for delivering results, as a matter of science politics. Our grant funders would demand results from our expedition. Making good on the stated mission mattered to the team, especially to me, as principal investigator—the lead scientist on the grant. Successful delivery meant many publications, ideally in the best journals. As someone just starting his career at a major institution, I felt added pressure to deliver. To my mind, that meant sticking to the plan and tying together a coherent story, not getting mired in too big of a side project with uncertain rewards.

But I couldn't unsee Cerro Ballena, nor did I want to. It glowed with significance, and it was the kind of unparalleled site that could boost the career of everyone associated with it (should it be successfully reported). Whatever explained the density and completeness of the skeletons was also surely an interesting scientific story, especially for understanding the evolution of whales in ocean ecosystems. But we needed to somehow wrap our minds around essential facts about the site before we could make any conclusions about how Cerro Ballena came to be or what it could tell us.

It seemed obvious that the sheer abundance of big whales at the site—clearly belonging to a species of extinct rorqual—was an important part of the puzzle. Getting more facts about those whales, however, would be daunting. A single whale skeleton might contain as many as two hundred bones, some as heavy as bowling balls or cumbersome as fallen tree limbs, topped off by a skull the size of a grand piano. Collecting a single whale skeleton, even one less complete than those at Cerro Ballena, requires coordinated effort by more than one person, usually over several days. First, you begin with what bone is available on the surface, and get on your hands and knees to inspect it—if it's not too fragile, there's no harm in carefully exposing it a bit more, by brushstroke, until you know what's there. Generally we try to save the time-consuming, detail-oriented work for the lab, but the bones can't get there unless they're exhumed from the ground. That means trenching a judicious perimeter around them and then wrapping them in plaster bandages that dry and form a hard cocoon. This protective shell allows us to securely pop the block containing the fossil out of the ground and bring it back to the lab for more work. Straightforward, but time-consuming when you're dealing with skeletal elements the size of furniture pieces.

When bones are scattered loosely all over the place, as is often the case, then the excavation becomes more like a crime scene, involving a grid (usually string hooked over nails or rebar) overlying the area to map out the position of each bone. The grid makes the relationships among all the bones clear and allows us to plot their locations on a sheet of paper. Then, after plotting, each bone is cataloged and removed individually, or sometimes collectively in small jackets, like the larger blocks. It's painstaking work.

In the case of Cerro Ballena, the costs of removing each fossil whale were the responsibility of the road-construction company. The company had agreed to pay Tuareg and his museum volunteers to expose, map, and document each skeleton to the best of their abilities, before jacketing the skeletons—sometimes whole, or in pieces as plaster jackets the size of tables and mid-size cars. At the largest sizes, plaster bandages and burlap would have failed, so metal support structures had to be added to ensure that the inner fossils were safely transported. While these steps saved the fossils themselves, we were losing crucial information every time a skeleton was pulled from the ground, separating the bones from the context—how they were buried, their orientation, their association with other bones, and so forth—that we would need to answer pressing questions about the origin of this site. The most pressing consideration of all: Tuareg said we had probably less than a month to study the site before all the fossils needed to be removed to make way for the highway's northbound lane. It was a stunning constraint.

With the specter of the skeletons at Cerro Ballena hanging over all of us, we measured and sampled all of the rock sections that we had originally sought. The next steps, in terms of assembling a single chronology of rocks in the Caldera Basin, would have to be largely lab- or museum-based. We had run out of time to do anything more. I knew we needed to act on Cerro Ballena. I just didn't know how. We barely had enough time to snap a few photos and measure the rocks at the site—nothing conclusive. It loomed in my mind, a whale-sized, multifaceted question that felt personal: What was I going to do about it?

On the return flight to the United States with Jim and Caro, I tried not to dwell on the logistics and instead thought about the bigger scientific questions. Was it a stranding? A one-time catastrophic event? An accumulation of bodies slowly over time? Testing any possible explanation for how all of these whale skeletons had arrived at the site would require an evaluation, bone by bone, skeleton by skeleton, to determine overall condition, arrangement, the articulation of each bone relative to the others, whether they had been scavenged, and so on. This type of forensic study would take time, definitely more than we had.

I wasn't sure which data were the most important to obtain fastest. The sheer magnitude of the task was overwhelming.

As I watched Jim ease into a window seat with his laptop for the long flight back to the United States from Santiago, I had an idea. If there were some way we could capture the bones digitally, beyond just static images, we might be able to buy time away from the site, back in the lab. I suddenly remembered a recent chance meeting with two people who did exactly that kind of thing.

I first encountered Adam Metallo and Vince Rossi in a windowless photography room in a research wing of the Smithsonian, hunched over the enormous skull of an extinct crocodile relative. They were passing a laser beam over its bony surface using a robotic arm connected to a laptop. Vince straightened to introduce himself, smoothed invisible creases in his untucked shirt, and launched into a demonstration. As he talked, I watched Adam paint light over the croc skull; the data then appeared in real time as a patchwork of surfaces stitched into a 3-D model on the computer screen.

Adam and Vince were part of a then-nascent 3-D digitization program at the Smithsonian, and they were looking for partners at museums throughout the Smithsonian constellation to provide case studies for their work. I thought what they were doing was genuinely interesting, but at the time I wondered what scientific problem scanning was solving. I thought it was more important to have a question first, and let the technology serve that aim. By the time I was thirty thousand feet over South America, poking my airplane-food portion of palate-blistering ravioli, I realized that I'd found that question.

Back at the museum, I did my best to present my quandary to Adam and Vince, convey my excitement, and hope for some kind of fruitful collaboration. In contrast with the traditional field methods in paleontology, it seemed obvious that laser scanning would have incredible advantages at Cerro Ballena. First and most crucially, it was many times faster than the time-consuming and costly steps of drawing grid maps or making large plaster molds with latex. Scanning would also have no physical contact with the fossil, eliminating any risk of damaging the bones when pulling off a latex peel. Second, scanning would create a digital model that was essentially a snapshot in time; there would be no decay of the facsimile, the way a latex peel or even a plastic cast slowly degrades over time. Given the scale of a single fossil whale skeleton, capturing a digital copy seemed like an ideal way to make comparisons across the potentially dozens of skeletons at Cerro Ballena—far easier to overlay or make side-by-side comparisons on a computer screen than pacing from skeleton to skeleton in the real world.

Laser scanning also surpassed the practicalities of other technological solutions. X-rays might perform similarly well, but there was no good way to bring an instrument that produced X-rays, such as a CT scanner, out to field localities. Nor was there any way to fit a thirty-foot whale skeleton in a standard hospital CT scanner. (Only a few CT scanners in the world can handle objects approaching a few feet in the largest dimension.) X-rays would have circumvented line-of-sight issues inherent for very large objects: the small overhangs and deep pockets along the vertebrae and skull pose the same challenges to a laser and camera as they do to a paintbrush full of silicone, because these nooks and crannies can't always be perfectly captured below the threshold of your application tool.

As it turned out, Cerro Ballena would be the first of many successful collaborative projects in the realm of 3-D digitizing of museum specimens. Adam, Vince, and I became fast friends; I immediately took a shine to their background in exhibit arts— artists and scientists aren't so different when you consider how creative enterprises see the light of day only when they squeeze past all the barriers of resources, personality, and time. Adam and Vince also knew that there were as many narratives as there were specimens in the Smithsonian's holdings, and that 3-D had the promise to make those stories accessible and visible.

Once I secured Adam and Vince's interest in my particular case, Caro and I went about securing funding for a last-minute field season with food, gas, trucks, lodging, and equipment— and airfare for now nearly a dozen people. We made a great team when it came to carefully calibrated funding pitches to the director's office at the museum, and to National Geographic, which funded part of our original season of fieldwork. As we flexed charm and rhetoric to underscore the urgency of the project, we managed to cobble together enough funds to cover the immediate costs of a return to Chile. In the flurry of activity, it became more and more clear that tackling Cerro Ballena meant moving out of a safe zone into the high-risk, high-reward category of scientific endeavors. For me, a pretenure curator, it was an unnerving consideration. Scientists earn credibility by finishing big projects and delivering results; spending a lot of time and money with nothing to show for it, especially on an international stage, means risking your future—and those of your early-career colleagues too. I didn't want to expose Caro, Tuareg, and Jim to those kinds of risks any more than myself.

When we returned to Chile, Tuareg was there to greet us; I introduced him to Vince and Adam, saving other concerns for later. We made directly for Cerro Ballena along the Pan-American Highway, arriving less than an hour after our plane had landed near Caldera, in Copiapó. With a week to go before Tuareg's deadline, we didn't have time for diversions. We needed laser-like focus.

Tuareg and his team had already made strategic decisions about which whale skeletons to document and remove and which ones to leave intact until we could arrive—less than a dozen, I guessed from a quick survey. In the brief interim, while we were gone, I knew that Tuareg had done the best he could to preserve what was at hand, as the road-construction company quickly scraped away the overlying rocks, down to the level of the highway. As an outsider, I had limited sway in Chilean politics; I wasn't sure what to expect until we rolled up to the quarries, which had been overlain by sediment only weeks before and now opened into a space about the size of two football fields end to end.

We had two basic approaches at our disposal, with very different logistics: a laser arm scanner (which required elaborate setup, especially with a heavy, shake-proof base to keep its position static during scanning) that collected the geometry of surfaces at submillimeter precision; and handheld, high-end digital cameras, which would collect images to build a 3-D model with software back in the United States. The laser arm scanner and its attendant gear were too cumbersome to move whale to whale and would need to be dedicated to a single skeleton; camera work, on the other hand, could be deployed across multiple skeletons.

One array of three skeletons piled on top of one another was clearly a high priority for photography: an aggregation people at the site had taken to calling La Familia. The two bigger, adult-sized skeletons lay tail to nose, splayed in a *V* but complete in every way, including even hip bones; joining them, in a small leg across the opening of the *V*, was a smaller skeleton belonging to a juvenile whale with unfused skull bones. I'm generally loath to use nicknames as a shorthand for individual specimens, mainly because I think that pet names trivialize the hard work that goes into collecting and studying fossils. At Cerro Ballena we used Tuareg's field numbers across the site, but "La Familia" was a lot more convenient than saying "B21, B22, and B23." I also had to admit that La Familia conveyed some awe for the oblong, heart-shaped outline of the trio of fossil whales.

There was no guidebook for the work that we set out to do, and it required a lot of thought, creativity, and trust. We camped out only yards away from the highway, and every eighteen-

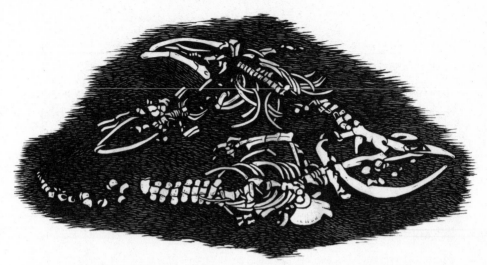

Clockwise: *B21, B23, and B22—or as they are better known, La Familia*

wheeler screaming downhill reminded us of that impending northbound lane. Tuareg and his team constructed a tent over B33, one of the most perfectly preserved of all the large baleen whales at the site, to shield the equipment and work from the dust and wind. We improvised seating on the largest transport cases, which also doubled as workbenches; Vince and Adam got priority with the only card table, which they used to create a workstation for their souped-up laptops. At one point I paused at the back of one of the field trucks and watched Adam and Vince busy at work with laptops, laser scanners, and tripods while cables and accessories spilled out around them. "You guys are like some kind of laser cowboys," I said, smiling. Adam grinned in reply and Vince laughed, but they both kept at work. The sound of all the busywork at Cerro Ballena gave me confidence, for the first time, that we might actually pull off this ambitious plan of turning the physical into the digital.

Days and nights melted away in the tent as the cowboys slowly moved from nose to tail, scanning B33. Each pass of the laser arm, like a thick brushstroke, was instantaneously transmitted to the laptop, revealing more and more of the skeleton's digital avatar on the screen before us. The fine resolution of the entire scan, highlighted in raking light by the computer program, looked like a bas-relief from antiquity. Nothing like it existed, as far as I knew—no fossil skeleton had been scanned in the ground with this kind of detail, and certainly not a fossil whale. I smiled for a moment, thinking how fantastic it was that we could capture and bottle the context of a skeleton like this one. There might be dissertations' worth of study in the data set from B33 alone. And we had also collected the photography data from all of the other whales, including La Familia.

Later, back at the Smithsonian, the cowboys would take these colossal, dense clouds of scanned data points and photographs and translate them into 3-D models of the whales at Cerro Ballena in incredible detail. The submillimeter precision of B33 (and the slightly coarser millimeter-scale precision of La Familia) was a way to future-proof the 3-D models: we collected information at the highest possible resolution and preserved each rendering at every step of the process. The basic underlying data were also recoverable well into the future by using standard formats for the digital photos and simple text files for x, y, and z coordinates of the scan data. These measures ensured that the time frame for studying the digital facsimiles of Cerro Ballena would extend for years to come.

The 3-D data from Cerro Ballena were mesmerizing to behold on a computer screen, far from the field site; but it's a whole other experience to 3-D print them, especially at the scale where they fit in your hand. Even today, when I pick up a color 3-D print of them—skeletons that took walking strides to appreciate and comprehend, now rendered compact in my hands—I'm transported back to our days at the site, even to the specific time of day, based on the shadows printed in relief. Even more exciting was the opportunity to 3-D print one of the Cerro Ballena skeletons at large scale, which today anyone can see at the Smithsonian, inside the natural history museum's learning center auditorium. Thousands of miles away and years removed from our work, I can walk the length of a plastic copy of B33 on a wall, seamlessly stitched together from individual tiles. I can tilt my head to just the right angle to see the skeleton, tinted with some paint, as it would have lain on the ground, and remember our long days beside the highway in the Atacama. These facsimiles tell stories as much as the real thing. More important, they can be shared with anyone who has access to e-mail, a computer screen, or a 3-D printer.

At the site, as Adam and Vince worked away at scanning, I shifted my focus toward the geological and paleontological data that we would need to test any idea about how this graveyard at Cerro Ballena came to be. During our first season, as part of our larger work in the Caldera Basin, we measured the stack of rocks exposed at the site. The stratigraphic column created from this work presents a kind of key that can be used to

determine the geologic age of the rocks. The column also provides a basis for interpreting the kind of environments that those
rocks represented—the overall setting—when the whales were
first deposited at Cerro Ballena.

By the second field season, we had realized something
profound about the site: Cerro Ballena wasn't just one layer of
whale skeletons but four, all directly on top of one another. It
took us some time to come around to this conclusion, in large
part because the Pan-American Highway cuts a long diagonal
gash against the horizontal rock units at the site as the road dips
to the south. As a consequence of the length of this slope, we
never had one complete view of all the layers at the site. From
the tent over B33, we could see other whale skeletons to the
north that appeared to be higher in the rock section by several
feet. But we didn't know if that was a real feature or the result
of the widespread faulting throughout the basin, which might
have broken up a single layer into a vertical jumble like uneven
blocks of a sidewalk.

A full reading of the layers at Cerro Ballena came only from
walking back and forth along the slope, moving up and down
in geologic time and tracing out individual layers from skeleton
to skeleton alongside the rock wall. This hands-on process let
us track the horizontal positions of individual whale skeletons
as we went, giving us visceral proof that there were indeed multiple layers of skeletons on top of one another. B33 was in the
second layer out of four; directly adjacent to B33, profiled on
the new road-cut wall, we could touch still more whale skeletons, unexposed, which were directly on top of it in the rock
sequence.

We also quickly realized that we needed Tuareg to help us reconstruct the placement of each whale skeleton in this new context—we were missing which whale came from what layer for everything collected before that insight. For all of his show-manship, Tuareg never failed us as a paleontologist, and we were able to deduce the information from his detailed notes and quarry diagrams. The fact that Cerro Ballena was actually four sites piled directly on top of one another was a stunning revelation for the scientists on our team. It was a piece of the puzzle hiding in plain sight, as important as the completeness or close proximity of each skeleton. Multiple whale-bearing rock layers at the site meant that there might be some unifying cause for why they were deposited in this one particular place—a basic proposition in science, in that repeated patterns tend to be generated by the same underlying process.

7.

CRACKING THE CASE OF CERRO BALLENA

I groped for my seat belt, sandwiched with three Chilean students along the bench of the Toyota pickup's cab. I couldn't quite grab the buckle before the truck suddenly took off in the dark, bouncing across a moonlit terrain way too fast. Tuareg erratically and enthusiastically tapped the gas. It was 10:00 p.m., we hadn't eaten dinner yet, and I had no idea where we were going.

It was our last night in the Atacama. All the geologic, rock-pick-in-hand work was done. The laser cowboys, close to finishing the tail end of B33, were scanning the perfect little disks of bone stacked in a sequence that comprised the midline, bony parts of the tail fluke. They had asked us to bring them back some food. ("Baked empanadas," Adam said. "Lots of them, please.") Sand-worn and ripe, we divided ourselves into two trucks to caravan to dinner. Caro and others went in one truck the usual way to our standard seaside spot, Punta Referencia.

Meanwhile, in Tuareg's truck, everyone was too busy talking to notice that Tuareg improvised a different route. We careened along what could barely be called a path, made by mining company trucks, across a boulder-strewn landscape. The whole truck rocked side to side as Tuareg rapidly gesticulated over the music blaring from the radio.

"Tuareg," I intoned. "Mario," I tried again with his real name. "Where are we going?" I knew that I was the stranger in a strange land, awkwardly wearing the hat of expedition lead. As the truck jostled, Tuareg paused in his story and looked at me in the rear-view mirror. "Ah, it is no problem," he declared confidently. "We're going to beat the other truck."

Tuareg returned to his story as we rounded a high point in the off-road terrain. I looked out at the moon-dappled waves of the Pacific Ocean in the distance. I could see, framed through my constantly shaking window, the outline of El Morro. It dominated the view for miles, a convenient landmark for sharing our location during fieldwork. It is what geologists call an igneous intrusion, the direct result of bubbles of deep, molten rock that ballooned toward the surface many millions of years ago, just now eroding down. Ultimately, subduction was responsible for El Morro, as it was for the stretch of exposed ancient seafloor where we were documenting forty-some whale skeletons. Although Darwin never wrote about El Morro, he could not have missed it on his way out of the Atacama. I imagined the *Beagle* tacking against the wind, heading to port in Caldera to collect an errant crew member who would change how we understand life on Earth.

I felt like I stood to contribute in a small way to that understanding by solving the biggest riddle of my professional life thus far: what happened at Cerro Ballena? It was an unusual opportunity because paleontologists generally have difficulty explaining unique events in the geologic past. Most of the time, we'll never know why any one particular thing happened—why *Pakicetus* went extinct, or why *Basilosaurus* had such a long tail. Instead, paleontologists are much more comfortable explaining classes of similar events, such as evolutionary convergence in form, or the relationship between the rise and fall of fossil species with changes in Earth's climate. At the outset, Cerro Ballena seemed like an isolated, single event, but on closer inspection, we discovered repetitive features—four whale-bearing bone layers stacked on top of one another—which made us think that there could be a unifying explanation about how the site came to be. In some ways, the riddle of Cerro Ballena seemed to have unique facts that begged a general cause.

For paleontologists looking to explain any site, the first step is often taken on the well-worn path of geology. In this case, the tan, loose sediments at Cerro Ballena reflected a single type of environment, which was most similar to a tidal flat. Each of the four layers containing skeleton after skeleton punctuated the otherwise constant backdrop of this environment—a tidal flat, in other words, was the persistent context for each one of the events that led to the bone layers. A tidal flat might not be an unusual place for a seal or four-limbed aquatic sloth, but it is certainly not where any whale, let alone dozens, would show up alive.

Judging from the thickness of rock at the site, and using an estimate for sediment accumulation in modern tidal flats, we calculated that the rocks encompassed a period of time roughly between ten thousand and sixteen thousand years. Though it's a range longer than the entire span of human civilization, it's a brief blip for a geologist. We didn't have any rocks that would yield precise radiometric dates, but the species of fossil sharks (based on a few isolated teeth) and aquatic sloths we found at Cerro Ballena were also known from rock units in Peru that were better dated. Using that correlation, we applied those dates to Cerro Ballena: our best guess was that Cerro Ballena was deposited over a ten-thousand- to sixteen-thousand-year window sometime between nine million and seven million years ago, during the late Miocene.

The large fossil baleen whales were the obvious stars of the site, being the largest and most complete skeletons at every bone-bearing level. All of these skeletons represented extinct species of rorquals, the gulp-feeding large whales with iconic living relatives such as humpback and blue whales. There was also a range of growth stages at the site, from yearlings to adults. Besides the extinct rorquals, there was a panoply of other creatures preserved at the site, including other marine mammals—something that gave not only a solid snapshot into the Humboldt Current during the Miocene but also a crucially important clue as to the origin of Cerro Ballena.

The Miocene lasted from about 23 million to 5 million years ago, a span that includes a fair amount of whale evolutionary history. Many of the continents were in their current configuration, although some ocean basins and portals between them existed where they do not today—North America and South America were cleanly separated, for example, by equatorial waters called the Central American Seaway. For long periods of time during the middle of the Miocene, including the time when the Sharktooth Hill bonebed was deposited, global sea level was higher than it is today, as were temperatures. These factors likely fostered the rich diversity we see in the fossil record of many groups during this time: whales were at their most diverse, occupying a range of niches and body sizes absent among their descendants today.

You can think of the Miocene as a kind of fever dream of the present, drawing on familiar members of today's ecosystems, in similar settings, but with occasional aberrant and nightmarish forms. If you went on a whale watch off the coast of Chile in the late Miocene, during the time of Cerro Ballena, you would have seen many whales that looked familiar: early relatives of today's rorquals and oceanic dolphins certainly plied the water, feeding on the same kind of prey that they do today. From a boat they might have looked largely the same, maybe different in markings and color patterns—though it's likely they had the countershading coloration common to most oceangoing animals (light on the bottom, dark from the top, to hide from their prey from both above and below). On closer inspection, the early rorqual relatives were not all that large (there were no blue whale–sized ones); and some of the dolphins may have had extremely long snouts, the way river dolphins do today.

But beyond the familiar, there would also have been killer sperm whales and walrus whales. Just yards away from B33 at Cerro Ballena, we collected a dozen banana-shaped teeth belonging to an extinct lineage of sperm whale that had both upper and lower teeth, covered in enamel. Today's sperm whales have only lower teeth; these miniature ivory anvils, which mostly consist of dentin and not enamel, seize cephalopods large and small. By contrast, their more toothy extinct relatives in the Miocene may have terrorized and consumed other marine mammals, including other whales.

At Cerro Ballena we also discovered tusks and parts of a skull belonging to a dolphin species with the face of a walrus: a foreshortened and downturned muzzle bearing two tusks, though strangely unequal in length. The only other fossils of this bizarre species have been found a thousand miles to the north, in Peru. Walruses today are restricted to the high latitudes of the Northern Hemisphere, and, to date, none of their fossils have been found anywhere other than the Northern Hemisphere. The fact that a lineage of toothed whale evolved a face and dentition nearly identical to a living walrus suggests that there was an analogous ecological opportunity off the coast of South America in the past few million years that persisted long enough for selection to generate that solution. (Walruses are separated by over 100 million years of evolution from whales.) Walrus-faced whales—*Odobenocetops* is their scientific name—likely fed on the seafloor in the shallows along the coast, in the same sea grasses that extinct aquatic sloths inhabited. (Yes, fossil sloths evolved long snouts and thickened ribs and limbs, all hallmark aquatic adaptations convergent with sea cows; and, like walrus whales, they have been found only in South America.)

There were also penguins and other seabirds in the Miocene, although some were very different from what's around today. Some fossil penguin species had much longer beaks, and one extinct seabird, called *Pelagornis*, had a seventeen-foot wingspan and cringe-inducing jagged pseudoteeth on its beak, typical

Odobenocetops scraping the Miocene seafloor for mollusks. Its long brittle tusks are asymmetric, with the right tusk growing significantly longer than the left.

of its now-extinct family of bony-toothed seabirds. As they do today, marine mammals and seabirds in the Miocene made a living feeding on the rich zooplankton of the Miocene Humboldt Current, although tracking their prey items is tricky without much of a fossil record. Krill, for example, have no fossil record, but analyses of clocklike sequences in the DNA of living organisms suggest that they have been around since the Cretaceous, some 100 million to 70 million years ago. It is a fair guess that krill were a part of the ancient Humboldt Current, but that's only an inference based on a similar lineup of predators that we see today.

One other group of predators also clearly played an outsized

Pelagornis, *the bony-toothed seabird of the Miocene*

role in the Miocene oceans: bus-length megatoothed sharks. These extinct shark species are known almost entirely from their fist-sized teeth, which are abundant in Miocene rocks the world over. Some species, such as the one commonly called "megalodon" (its proper, scientific name is still a matter of debate among specialists), grew forty feet long and ate whales, among other marine animals. No shark scavenging marks were found on any skeletons at Cerro Ballena; the few shark teeth we did find belonged to extinct species smaller than "megalodon," and probably washed into the tidal flat or were left by intermittent scavenging episodes at high tides. Some scientists have argued the ecological presence of these large sharks elsewhere prevented baleen whales from evolving the large body sizes that we see today; the timing of their extinction, around three million years ago, coincides with the rise of extremely large whales, although the true cause of the extinction of these terrifying sharks remains mysterious.

We tallied a total of ten different kinds of large marine animals at Cerro Ballena—all creatures at the top of the food chain, including both carnivores and herbivores. It was a peculiar list: whales of many kinds, multiple seal species, aquatic sloths, and billfishes. And, most amazing to me, our inventory captured all of the unique fossil marine mammal species that have been found only in South America (walrus whales and aquatic sloths), within just a few hundred yards of road cut. That lineup told us an important clue: there was likely some kind of factor that concentrated them all there, in one spot, over the course of at least four separate episodes.

Knowing the geological stage and the ecological players narrowed the explanations for the origin of Cerro Ballena. How was it that both big, oceangoing marine carnivores and herbivores ended up feet and yards away from one another, some perfectly preserved, stem to stern, like B33? How did air-breathing and gilled vertebrates (namely the billfishes) end up in the same site? Last, and most important: how could all of these factors line up four times in the same place?

Instances when different species of cetaceans or marine mammals washed up together were rare, but they were well documented, especially in the United States. And interestingly, every time a multispecies stranding had a putative explanation, harmful algal blooms were the common (and sole) culprit. HAB for short, these single-celled organisms—including different species of diatoms, dinoflagellates, or blue-green algae—have the ability to cause widespread death in marine consumers because the toxins that they produce become concentrated as larger and larger organisms higher up the food chain eat more and more of them. While HABs include a wide range of species and toxins that manifest themselves in a variety of ways (algal blooms, red tides, and the like), they all work the same way: when the single-celled organisms reach sufficiently dense aggregations—a so-called bloom, at a local scale—its toxic effects become magnified as bigger and bigger consumers eat more and more poisoned prey. Ultimately something microscopic can kill the top players in marine food webs—whales, other marine mammals, bill-fishes, and humans too.

If the HAB explanation for Cerro Ballena was right, it also explained an important feature we had seen in the geology of the site: orange halos throughout each bone-bearing layer, which our team's sedimentologist attributed to iron staining from algal mats. While it would be difficult to determine if the organisms that caused these formations were *the* algae of death, it was clear that they fixed iron that was mobile in the environment. Iron is extremely important where HABs are concerned: in the modern world, iron supercharges their potency, increasing the extent and duration of a bloom.

Science works by eliminating bad explanations until we're left with the simplest argument (or the fewest steps) with the broadest explanatory power. No other explanation for the peculiarities of Cerro Ballena held up to scrutiny. It wasn't a group whalefall—far too shallow. If you wanted to be fanciful, you might think that an earthquake or a tsunami might somehow send a raft of carcasses into a tidal flat. Not only are there no modern examples to support this idea, but also the geology of the site showed no such high-energy, catastrophic sedimentary structures. And most important, any death mechanism would have needed to have happened in more or less the exact same way (and same place) four times. Where other explanations failed, or only explained a particular set of observations, HABs could explain the totality of what we knew about Cerro Ballena.

As best I can tell, the story of what happened at Cerro Ballena goes something like this. Sometime in the late Miocene, between nine million and seven million years ago, a vast, shallow tidal flat opened to the South Pacific Ocean, perhaps slightly corralled by a small chain of rocky islands and surrounded by a hyperarid desert inland. Offshore, now-extinct species of whales, bizarre dolphins with walrus faces, ancient seals, and other marine predators thrived in the Humboldt Current, in a way perhaps not that different from today.

Then, iron runoff to the coast from nearby rivers fueled widespread and toxic algal blooms that killed legions of marine mammals of all stripes and ages. And they killed these top consumers rather rapidly: perhaps within hours, through asphyxiation—the awful way that algal poisons suddenly kill whales in the modern world. Seals, sloths, and even billfishes were not immune to the widespread effects of algal neurotoxicity either. Wave action and currents—sometimes even storm surges—pushed carcasses close to shore, over the course of days to weeks. The carcasses of the largest whales arrived mostly on their backs, belly up, except when waves occasionally turned a few over. Ending up in the northern corner of the broad tidal flat, the carcasses were concentrated, like a logjam on a river, until the tide passed out, stranding the carcasses on the flat. The three whales that made up La Familia were among probably hundreds of others; they were likely related to one another, at a population level, and were buried as they lived, together. They mostly rested intact, flesh scavenged by saw-toothed seabirds and crabs until they were buried slowly. This sequence of events happened over the course of at least four different episodes several thousands of years apart, each time ending in the exact same spot, producing a stack of skeletons that would later become Cerro Ballena.

Over the course of millions of years, the stack of rocks from the ancient tidal flats were uplifted by tectonic activity, rising nearly two hundred feet above sea level. During this time, the forces of compaction and chemical replacement flattened the skeletons and turned them to rock, especially with weight of the overlying strata—an aggregate of shells called a coquina—deposited during the ice ages. This whole stack of rocks rested on the northern edge of the Caldera Basin, overlooking Caldera and El Morro, to the south. It would take until the late twentieth century for the first road cuts to blaze a path through these strata, destroying any number of skeletons, and then until the twenty-first century for our story to take place, measuring, scanning, and cataloging all of the bones for removal and their digital legacy.

I originally went to Chile to understand the material ways that past ecosystems were different from the oceans of today—in essence, did walrus whales, aquatic sloths, and bony-toothed seabirds matter? I also wanted to know how the differences between ancient and modern species arose over geologic time, which chronicles the rise of whales, and other large consumers, as important ecological players. What we discovered, through the long but ultimately valuable diversion of Cerro Ballena, was that dense concentrations of bones can also provide some of the most telling ecological snapshots of these systems. In two football fields of highway road cut, we sampled a complete list of nearly all known marine mammal species in South America, from walrus whales to aquatic sloths. It would not be surprising if we found other species known from elsewhere in Chile and Peru, should anyone ever excavate farther into the side walls of the road cut. By our estimates, there is another million square meters of rock at the site with that preservation potential, amounting to hundreds of possible Las Familias of walrus whales, seals, and other species.

The story of Cerro Ballena is probably not unique in the whole of Chile, nor even South America. Anytime we see recurring events in the geologic record, it tells us something important about broader processes in Earth systems. If our argument about the repeated mass strandings at Cerro Ballena is true, then there should be fossil whale highways elsewhere in the world: anywhere, at least, where the traces of upwelling zones get preserved by plate tectonics, along with embayments, tidal flats, and killer algae. Specifically, we can look to westerly continental coastlines around the world, where there are pockets of an accessible fossil record and the right environments preserved in the rocks. Geopolitics was the impetus for the discovery of

Cerro Ballena—it may take this kind of serendipity to find others as well. And what should be true for whale mass strandings should apply equally to mosasaurs, ichthyosaurs, or any other kind of oceanic consumer in the past, given the deep antiquity of single-celled organisms that have been causing harmful algal blooms for nearly a billion years.

While most other marine mammal bonebeds are the result of prolonged processes, preserving bones on the seafloor, Cerro Ballena was notably different: a tidal flat that preserved loads of carcasses that were physically concentrated by wave and storm action and preserved in dense intervals. This density of individual fossil marine mammal skeletons found at Cerro Ballena is unrivaled by any other site in the world, whether it's Sharktooth Hill or even Wadi Al-Hitan. I've always thought that the window Cerro Ballena reveals about the Humboldt Current in Deep Time puts it on par with other UNESCO World Heritage sites (including Wadi Al-Hitan). And we're fortunate that the accessible digital legacy that we built is available right now to anyone with an Internet connection, because there is still so much to learn.

East of El Morro the foothills give rise to the Andes, which form the spine of the entire South American continent. They are a dramatic consequence of the Nazca Plate's long-term plunge beneath the greater continental South American Plate, pushing strata to elevation the way pages of a thick book roll in a soft cover. The rocks in this part of the Andes are much older than Cerro Ballena—some as old as or older than the first dinosaurs.

In this region, not more than a day's drive away from Cerro Ballena, sits another result of the collision between geopolitics and science. Hyperarid desert air, the highest elevations in the Americas, and locations far from urban centers of light pollution make for the perfect place to mount the world's most advanced astronomical observatories. The Smithsonian's first solar observatories were also based here, built at the turn of the twentieth century, long since surpassed by fortresses containing telescopes that peer billions of years into the universe's past. Astronomy and paleontology are sibling fields, really: they take human imagination to places where no person has ever been, and they are endeavors entirely open to contributions from anyone, so long as you have the persistence to join the hunt. Any day an amateur might find a comet or a fossil that forever changes textbooks. It's happened before, and it will happen again. Astronomers even hijack the word "fossil" as an analogy for the light from distant stars, novae, and galaxies, a diverse bestiary of phenomena not unlike the assemblages of extinct forms buried (and occasionally exhumed) from rock. We both need data, whether from time on the telescope or hands and knees on the ground, scraping for bones.

When our truck finally pulled up to Punta Referencia, raucous cheers of excitement greeted our lagging team. Tuareg sauntered in to a hero's welcome of hugs and kisses on the cheek. I held back near the truck, away from the warm glow, to take a quick peek at the clearing night sky. I was hoping to catch sight of the Southern Cross, something I could never do at home.

I walked toward the beach, past two dogs tussling in the shadows, until the alleyway gave way to a full view of the sky's blue and purple dome. Only a few hundred light-years away, most of the stars in the four-pronged kite of the Southern Cross had been glowing for at least ten million years before dozens of Miocene whales washed up at Cerro Ballena. I thought about how the starlight above and the fossils beneath our feet existed long before the first human tribes craned their heads to the night sky and will persist long after the last humans become extinct. Against that, in the here and now, we hunt for real things about the past—pixels of light or worn bones—that give us a material way to interact with scales of time far outside our own.

Famished, and a bit lonely, I returned to the restaurant, tucking in shoulder to shoulder with students shivering in fleece and nylon jackets. But the cool evening air hardly mattered, as the beer and jokes flowed, along with freshly fried seafood spilling across the wood table. Many of my Chilean colleagues were children of the dictatorship years; they had no real intellectual continuity with their predecessors, yet they succeeded in finding their way. The ringleader, Tuareg, held court, drawing extinct, taxonomically appropriate centaurs of everyone on bar napkins: Caro was a long-beaked river dolphin; I was a walrus whale. One of Caro's students leaned toward me. "This is classic Tuareg. You will find cartoons like these on napkins in every cafe from Antofagasta to Punta Arenas." I believed it. I imagined Tuareg's trail of cartoon napkins and cell phones across the skinny spine of Chile.

————————

In the end, science is not a solitary endeavor. It takes partner-
ship, whether it's massive teams to ply the night skies with telescopes on mountaintops or legion ship crews to carry naturalists
across many oceans. Or paleontologists gathered at a bright,
friendly spot off the fossil whale highway. Tuareg deserved full
credit for first realizing the scope of Cerro Ballena and persisting through my doubts; I was happy to direct the digital salvage
of the site and pull together the evidence we needed to figure out
how Cerro Ballena came to be. With my seat at the table, I felt
lucky to be a part of it all and, in that moment at Punta Referencia, didn't feel especially much like a foreigner. Instead, I felt
like just another fossil hound, gathered at the evening's camp-
fire to share news of the day's discoveries and a bit of gossip.

PART II
PRESENT

8.

THE AGE OF GIANTS

Take in a deep breath. And now relax. You just shared oxygen with the largest animal—ever, in the history of life on Earth. In a single blow, this animal expels a column of water vapor that would reach the top of a two-story house; the air that passed through its lungs is enough to fill half a cement-mixer truck. Its blood cells pass through a heart with vessels the diameter of dinner plates and run through a body with over twenty billion miles of arteries, capillaries, and veins. Blood must carry oxygen to every cell in this animal—all 1,000 trillion of them—including nerves whose fibers reach over one hundred feet from its brain stem to every extremity, including its tail fluke. With a flip of its fluke, the animal descends to a depth beyond the limits of light, where it makes the most acoustically powerful sound made by any organism, a low tone that spreads for over nine hundred miles, echoing off undersea canyons. Everything about this animal, in all the ways that we can measure, is superlative. This animal is, of course, a blue whale.

Extremes enthrall us all, and the idea of an ultimate size champion among living things is a captivating thought. Ask an elementary-school kid about the biggest animal of all time, and the short list of possible contenders very quickly becomes a battle between two titans: a blue whale versus a sauropod dinosaur. This comparison is enshrined in textbooks, usually with a stylized composition of a blue whale, midgulp, floating above a sauropod, sloping neck and tail extending in opposite directions. Sometimes you might see a string of African elephants or school buses for scale.

The winner of this runoff, of course, depends a bit on how you measure the contestants. The longest of the sauropod dinosaurs probably exceeded 110 feet in total length, based on relatively complete skeletons. That linear distance comes close to the longest blue whale ever measured—a 109-foot female from the Southern Ocean, killed in 1926 by Norwegian whalers. But whales are the true heavyweights. At most, the largest sauropods might have weighed 120 tons, but the best estimates place the largest ones closer to 70 tons. By comparison, the heaviest blue whale reliably measured (a female, also from the Southern Ocean) weighed 136.4 tons, or just over 300,000 pounds— more than the takeoff weight of a Boeing 757. And this particular whale was only an eighty-nine-foot-long female. A blue whale closer to one hundred feet, especially if pregnant, would have weighed much more. From the standpoint of biomass, it's not much of a contest: blue whales are the most massive animals ever in the history of life, and we just happen to live alongside them.

For us, a whale's size may be the most striking thing about them, but the fascination is not just about sheer bulk. How can a creature of that magnitude keep itself alive? Size determines a lot about how much air a whale needs, how deep it can dive, how much food it needs to eat, and how far it can swim. But step back for a moment from these physiological questions: how did whales become giants in the first place? We know the beginning and end points: within the fifty million years that took whales from land ancestry to dominance in the oceans, the weight difference between *Pakicetus* and a blue whale increased about ten thousand times. How exactly did this change unfold? And how can we know?

Those questions loomed over me for many years. As a graduate student, I began to chip away at them, starting with the question of whether the roles extinct whales played in past ecosystems were similar to the niches they filled today, or if those ancient roles had since become obsolete. As a student, I had already read enough about walrus whales, for instance, to know that there were some cast members of past whale worlds that lacked any descendants.

I also spent a lot of time measuring whale skulls, keeping a log of the same set of measurements for every specimen I encountered. I knew the more data points I had, the more useful the information would be in helping me (and others) predict the total length of any fossil whale when only the skull was available. (As mentioned previously, whole skeletons, nose to tail, are essentially never recovered, Cerro Ballena notwithstanding.) I came to realize something odd about the fossil record of both baleen whales and toothed whales as well: none of them were as big as even their middling descendants today.

I hadn't deeply probed this surmise with analytical tools. At the time, I was more focused on the challenge of reconstructing a single whale's body size given scanty inputs. But I knew there was something to my observation. At Wadi Al-Hitan in Egypt, perhaps the best snapshot into the life of whales during the late Eocene, *Basilosaurus* was the whale titan; yet it was only as long as today's humpback whales (a midsize baleen whale at about fifty feet long) and weighed probably five to six times less with its comparatively much smaller head, short chest cavity, and long tail. At Sharktooth Hill in California, the largest whales were even shorter, no more than thirty or so feet long. While those Miocene whales had modern proportions, none were nearly as large as a modern humpback. Even for all of the millions of bone shards in the Sharktooth Hill bonebed, not a single one represented a fossil whale species that remotely approached the gigantic sizes of today's blue whales.

I didn't know what to make of the broad observation and repeatedly pushed the issue back under other more immediate layers of inquiry. But the question always resurfaced. Years later, at the Smithsonian, I came across a postdoctoral research fellow who could finally help me get to the answer. Now a professor, Graham Slater specializes in analyzing evolutionary trends in the fossil record. Using my working file of skull measurements, we charted the sizes of all living and fossil species across their family tree; we focused particularly on baleen whales (excluding toothed whales, which feed at a variety of different water depths, by echolocating on prey). We finally had an evolutionary map for how size evolved in whales over the past thirty or so million years.

Our analysis revealed that very large body size in baleen whales had evolved several times across different lineages. Extreme gigantism in whales—body lengths of more than sixty feet or weights over 200,000 pounds—appeared on different parts of the tree in species unrelated to one another. Blue whales and fin whales, the longest and second-longest whales today, for example, are actually not very closely related to each other. And neither of these whales is particularly closely related to right and bowhead whales, which can, at their maximum sizes, also tip the scales at one hundred tons. When a notable feature arises several times on distantly related branches of an evolutionary tree, it's a sure sign that something interesting is going on.

Our evolutionary tree mapping also confirmed another fact I knew casually: the world today is devoid of very small baleen whales, compared with other times in the geologic past. Many of the fossil baleen whales that Remington Kellogg described in detail, including those from Sharktooth Hill, were very small. In some cases, their skulls were small enough to cradle in your arms. The total length of these filter-feeding whales would have been about the size of a car—similar in size or smaller than today's enigmatic pygmy right whale, the smallest baleen whale. In the past fifteen million years or so, whales at this size seemed relatively common in the fossil record; our analysis showed that they went extinct very recently, some species only a few million years ago.

Past whale worlds were very different, based on the story of size alone. At this evolutionary moment, we live in an age of giants, sharing our planet with the largest whales ever, some of which are the largest animals to ever exist, period, in the history of life. There's nothing the size of today's blue whales, fin whales, or any of the skim-feeding whales in the fossil record. The immediate questions then become these ones: What drove the history of these changes in the fossil record of baleen whales? Does it have more to do with the whales themselves or the worlds they lived in? And what's preventing them from getting even bigger?

You might expect baleen to hold some answers. Baleen whales are, after all, the largest of all whales, and they have a unique anatomical apparatus that makes them stand apart from all other mammals, alive or extinct. Look inside the mouth of a whale, and you'll know immediately which of the two major living groups it belongs to: if it has rows and rows of plasticlike plates hanging from the roof of its mouth, it's a baleen whale. If it doesn't, it's a toothed whale. Toothed whales all echolocate, and despite their name, a few species don't actually have teeth, as they hardly need to chew or break up their food after seizing or sucking it into their mouths.

Baleen is a soft, pliable structure made up of keratin, just like fingernails, hooves, and hair. It grows from the roof of the mouth, soon after a whale's birth, emerging in a series of plates, numbering in the hundreds, which form a racklike structure. These triangular plates are ensconced in a bed of flesh, are fed by blood vessels and nerves, and grow in layers the way our own nails and hair do. Baleen develops frayed edges as its tubules are worn down from the friction of feeding; these tubules spring outward, like uncooked spaghetti, interdigitating with tubules from adjacent baleen plates to form a mesh facing the inside of the mouth. When baleen whales close their mouths around a gulp of water laden with krill, fish, or other zooplankton, the racks of baleen create a filter that traps prey inside.

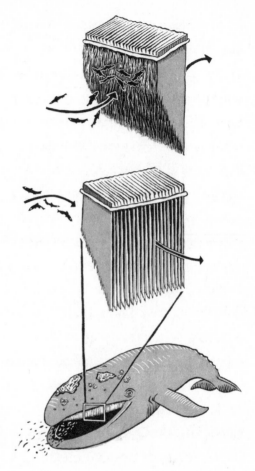

Baleen is the filter for filter feeders

So how long has baleen been around, and does its appearance relate to whale gigantism? It turns out that the first baleen whales didn't have baleen, meaning, yes, they once had teeth. There are three major pieces of evidence that underlie this point. First, modern baleen whale fetuses have tooth buds, primordial bits of tooth enamel and dentin resorbed by the body prior to birth. Baleen itself grows from ridges on the roof of the mouth during the first year of life, but in the womb the genetic machinery for creating teeth is still in place, a vestige of baleen whales' deep past as much as tiny hind limbs in *Basilosaurus*.

Second, baleen whales share common ancestry with toothed whales; these two branches that comprise the surviving houses of the whale family tree diverged around 35 million years ago, during the last days of *Basilosaurus*. The first baleen whales hardly looked anything like a blue whale; they looked much more like an extremely shortened version of *Basilosaurus*, with tiny versions of jagged, saw-blade teeth lining their mouth. They weren't eel-like in body proportions but built more along the lines of a bottlenose dolphin. But we know that these first, toothed members of the lineage that eventually led to filter-feeding whales are more closely related to blue whales than they are to any dolphin, because they share key traits located in their unique, pebblelike ear bones.

Third, although baleen is a soft-tissue structure, it does occasionally show up in the fossil record. Its preservation may have to do with a unique chemical environment on the seafloor. Fossil baleen has been reported from rocks in California and Peru, although no more than about fifteen million years in age—about half as old as the oldest baleen whales.

One important caveat is how we define whale groups when talking about baleen whales: because the earliest baleen whales didn't have baleen, describing these extinct forerunners using common words is challenging. Scientific nomenclature helps— "mysticetes" instead of "baleen whales," "odontocetes" instead of "toothed whales." At some point on the separate branches leading to today's baleen whales and toothed whales, filter feeding and echolocation evolved, respectively. But we don't know exactly when, how, or how many distinct times. Consider birds and dinosaurs: when scientists discovered dinosaurs with feathers, it became clear that feathers do not necessarily make a bird. Instead, we need to define organisms as lineages separate from traits that diagnose their identity.

Some scientists have proposed that early baleen whales had a combination of teeth and primordial baleen—teeth outboard, baleen inboard—but fossilized primordial baleen hasn't been found in any of the species where such a scenario might be expected. Other scientists have proposed that the first baleen whales evolved a half-and-half solution, but oriented not outside-in as in the other proposal, but front to back: front teeth still in place but lacking them in the back half of their mouth, instead bearing a primitive ridge of baleen. Still another possibility is an intermediate state with neither teeth nor baleen—simply a toothless, gulping early baleen whale. Most of these scenarios involve interpreting fragmentary fossils that don't preserve the critical soft-tissue structures; no one has yet, for instance, demonstrated how a whale with both outboard teeth *and* inboard baleen would actually be able to feed. Absent more fossils with clear evidence for what the first baleen plates looked like, all that we can do with these ideas is to test them using biomechanical or computer models. Until scientists figure out a way to record video inside the mouths of today's filter-feeding whales— a logistical hurdle that may be impossible to surmount—we need to shore up our knowledge about how filter feeding works in living baleen whales before jumping to any conclusions about how it may have worked in their ancient fossil relatives.

At the Smithsonian I can place all of the skulls and jawbones from the first baleen whales from one geologic epoch on a single long table. The skulls of many of these first baleen whales— from a time called the Oligocene, around thirty million to twenty million years ago—have prominent large eye sockets and flat, beaklike snouts, with sockets for teeth. Some of these fossils preserve teeth in place, which makes me wonder about all the ways we could test their feeding style. The chocolate and coffee tones of the bones contrast against the gray mud-and-siltstone matrix that encases them. The rock matrix for bones this ancient can be so hard as to require years of careful exhumation by pneumatic chisel, acid baths, and the sharp picks of dental tools.

Many of these Oligocene whales hail from the storm-washed beaches of the Pacific Northwest and adjoining sea cliffs. Amateur scientists are responsible for nearly all of the Smithsonian's collection of fossils from this part of the world. For these fossil finders, the brutal weather can make the finds more satisfying: winter storms wash the shores with heavy waves that pull away sand to reveal boulder after boulder, many with bits of bone inside if you know how to read the rocks. And, as always, you need to be lucky.

The ear bones among these Oligocene fossils reveal their identity as close forerunners of today's baleen whales. Otherwise their needlelike and slightly cusped teeth recall some kind

of strange, monstrous seal, or might make you think they belong in the Cretaceous, along with marine reptiles in the time
of dinosaurs. But they are indeed whales, albeit smaller versions
of today's giants transposed tens of million years into the geologic past.

Their descendants, still alive in today's oceans, have skeletons that vastly eclipse these early whales in size. Baleen came
first, then, much later, enormous sizes. Even at Cerro Ballena, a
site that samples the diversity of whales from nine to seven million years ago, no single baleen whale skeleton reaches more
than about thirty feet in length. The origin of baleen doesn't
neatly explain why blue whales and other filter-feeding whales
became modern-day giants. We have to keep digging.

9.

THE OCEAN'S UTMOST BONES

There is a set of unassuming warehouses in Prince George's County, Maryland, which holds secrets of immeasurable value. The corrugated metal sheds stand at the end of a nondescript asphalt road, where the Smithsonian keeps its off-site collections, including Apollo-era spacecraft and planes from both world wars. They also house the focus of my life's work and that of many others: thousands upon thousands of whale bones from nearly every species alive today, and many long extinct. These vaults hold an archive of past whale worlds, ranging from geologic to historical times, that has no equal anywhere else.

Once inside, the smell is the first thing you'll notice: a latent oil odor hangs in the air, still emanating from century-old skeletons. Rows of steel shelves hold the full series of vertebrae from the largest whales—the neck, chest, lumbar, and tail bones of blue whales, fin whales, right whales, and sperm whales—all stacked like giant dominoes and wedged in archival foam. The natural oil that still lingers in these skeletons smells like a thousand wax candles charged with smoke and seaweed. For me, this dense and heady smell conjures memories and associations about these bones that have formed over years of study. When I notice the smell laced in clothes or on my hands, it reminds me that, for any question about whales, there is probably a specimen in this collection that could answer it.

Metal frames line the floor, ten to twenty feet tall, each with skulls and jaws lashed to its structure, mounted upright. Casters allow one person to move the enormous specimens with comparative ease. These specimens, and the thousands of bones housed in cabinets and resting on shelving, represent a physical record of nearly every whale species alive today, including some yet to be named. (Yes, there are still new mammal species to be formally named in the twenty-first century.) A large portion of the fossil whale collection at the Smithsonian is stored here, providing perhaps the only place in the world where the skulls of a blue whale, a fin whale, and a number of other gigantic whale species can be compared side by side with those of their long-extinct fossil relatives. With vertebrae the size of desks, jaws as large as telephone poles, and skulls that wouldn't fit inside a city apartment, the chain of human effort to bring just one of these items to the museum from its point of discovery is almost too exhausting to contemplate. Multiply that single custody chain by tens of thousands of specimens over 150 years, and the cumulative effort is staggering.

There is one whale specimen, however, that stands apart from the rest. USNM 268731 is the catalog number for the right and left jawbones of the largest blue whale specimen in any museum in the world—I've looked for and measured all other possible contenders. Nearly twenty-three feet long, each weighing about a ton dry weight, the bones rest horizontally on the largest custom metal frames in the collection. Neither would slide easily through any kind of normal front door; it takes half a dozen strides and a few breaths just to walk their length. Bladelike terminations mark where the two would meet to form the whale's chin. (Our own right and left jaws are fused at the chin as a bony symphysis by the time we are born.) At the top of each jawbone is a fine peak, called the coronoid process, which provides an attachment, in life, for a set of straplike muscles that pull the jaw toward the skull. Two large bumps nearby, on the other end, anchor the jaws into the skull itself.

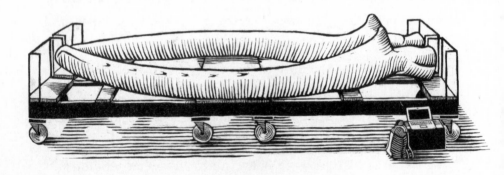

USNM 268731

Bones this large seem improbable; it is hard to wrap your mind around something the height of a football field goalpost attached to a moving, breathing being. In all of their fundamental dimensions—length, width, height, and weight—these bones are larger than the biggest mammoth tusks or the bones belonging to the largest dinosaurs. Blue whale jawbones are not just the ocean's utmost bones (to borrow from Melville) but the utmost bones in the history of life on Earth. Of course, they did not just appear in the museum, as if fabricated out of some incredible special-effects studio. They have an origin story, like every other specimen in the museum's collection. And in this case the story begins on an island in the middle of the Southern Ocean, about a hundred years ago.

The Drake Passage, a stretch of churning, frigid, and perilous seas, stretches between the southern tip of South America and the craggy archipelago of the western Antarctic Peninsula. It formed some thirty million years ago as the tectonic plates of South America and Antarctica separated, finally giving way to a continuous current system that now rings Antarctica. The Circum-Antarctic Current is unimpeded by any landform, allowing wind and waves to build incredible force, keeping albatrosses aloft and menacing navigators for centuries. This current also keeps Antarctica refrigerated and fuels the rich ecosystems that support krill, whales, seals, and penguins.

For those fortunate (or crazy) enough to cross the Drake, the immense heave and pull that you feel under your feet, or the towering waves that rattle even the sturdiest vessel, remind you that we live on a planet that frequently defies our attempts to control it. After an ill-fated expedition to cross Antarctica on foot in 1914, the explorer Sir Ernest Shackleton and his crew were marooned on an island at the tip of the Antarctic Peninsula. Crossing the Drake in open boats would've been folly; instead, he and his crew aimed for South Georgia, nearly eight hundred miles away to the east—with the current, not against it. In a feat of survival essentially unparalleled in the modern age, Shackleton landed with his men on the south side of the island before making the first-ever inland crossing of the glaciated island. Exhausted but unbowed, Shackleton sought a place on the east side where he knew he could find help: a whaling station.

It's as unlikely a place as you might imagine for one whaling station, let alone the several that existed at South Georgia in the early twentieth century. At the time, each was more like a small port town, with barracks, administrative offices, churches, and, of course, factories specially designed to pull whole whale carcasses out of the water for rendering. The names for these stations conveyed claims from nations a world away—Stromness, Prince Olav, and Grytviken—and operated under companies with different flags, including Norwegian, British, and Argentine. Today these abandoned whaling stations are protected heritage sites under British territorial claim. But in the early twentieth century these ports were filled with smells and sights from the most massive sustained whale hunts in history, pushing many whale species in the Southern Hemisphere to the edge of extinction.

South Georgia's heyday was brief, elapsing between two technological innovations: in 1864 the deck-mounted explosive harpoon made whaling much more lethally efficient, and in the late 1920s gigantic factory ships freed whalers from the need to process their catch on land. As a seamount, the island's underwater topography created a local upwelling that naturally attracted an abundance of whales nearby—in the thousands, so many that whalers could hear their blows echoing throughout the harbor. Lacking any kind of restrictions, whalers devastated the locally abundant whales and then pursued whales farther afield in the great Southern Ocean. At its end, Southern Ocean whaling in the twentieth century accounted for over two million of the more than three million whales killed last century.

The scope of this undertaking, which began on a remote island, is difficult for us to grasp today, but it rivals more widely
known exterminations on land, like those of the American
bison and the passenger pigeon. Shackleton's photographer,
Frank Hurley, left a legacy of images collected on the expedition's initial leg southward, showing Grytviken with corrugated
tin sheds next to long, wide flensing platforms, like airport tar-
macs, tilted into the harbor. Ice fields and jagged hillsides rise
dramatically in the background. The enormous whale carcasses
in the foreground are startling: in one photo, a nearly hundred-
foot-long blue whale dwarfs the gaggle of men beside it. Few
people alive today, if any, can relate to the sight of a carcass that
massive. While only about 150 blue whales were ever killed at
these lengths, over 325,000 blue whales of all sizes were killed
during Southern Ocean whaling in the twentieth century; today blue whales are a rare sight in these waters. It's quite possible that the gigantic, limit-pushing blue whales have had their
genes removed from the population by whaling. At the least,
it will take a few more decades for any surviving calves from
that era, now fully mature adults, to reach the lengths of their
ancestors.

While commercial whaling took a colossal and unprecedented
toll on many species, it also provided unique information, creating an uneasy intersection between interests. Whaling—across
geographic and temporal scopes—has given us a lot of basic
biology about whales that we would not know otherwise. The
sheer scale of industrial whaling was a lethal way to sample

whale diversity across the world's oceans. The tabulations and maps generated out of this killing provide some estimation of what whales lived in which oceans at a specific time. Very few of these physical vouchers ever made it to a museum; whaling in the twentieth century was about profit, in the form of oil and meat. But the occurrence data for each recorded kill—which species, where it was harpooned, and when, like a waypoint—also provide an account of whale biodiversity in the early twentieth century that we will never get again. (We do know that these data aren't entirely unbiased. Soviet whalers, especially in the North Pacific, consistently underreported their whaling catch statistics for decades.)

Whaling has also provided us with volumes of anatomical data. For example, our sources for total body weights, organ weights, length data, and relative timing of age to physical maturity for a variety of large whale species is derived from the whaling industry. There are also details about reproduction (birth rate and gestation based on ovary data, for example) and diet (based on gut contents—a last meal), which permit the identification of prey species in a precise and quantitative way. At the height of whaling at South Georgia in the 1920s, many of the size measurements (such as total length or girth at the pectoral flippers) became standardized, giving this kind of reporting comparability and repeatability. Sometimes even data about parasites, outside and inside the body, were collected.

To support an industry of empire, the British government formalized data collection about whale landings when it convened the Discovery Investigations from 1918 to 1951. The output of this scientific inquiry into Antarctic whaling—organizing

and analyzing whale data and their oceanographic context—
weighed in at thirty-seven volumes and was made publicly available. The *Discovery Reports* are scientific proceedings, but they
also reveal the immense scale of the logistics involved in the
hunts and the hardships of living and working at sea or on a
remote island. When I page through the *Discovery Reports*,
full of tabulations, sepia photos, and line drawings, I wonder
whether the scientists who collected these data recognized that
the chance to do so would never happen again.

One person who would have pored over the *Reports* at the time
of their publication was my direct predecessor at the Smithsonian, Remington Kellogg. I inhabit a world he created: as the
current curator of fossil marine mammals at the Smithsonian, I
am the steward of the collections that he largely built. It is the
largest collection of its kind in the world (by a fair margin, measured both in sheer tonnage and in variety of extinct branches
of the whale family tree). Much of it was collected, handled,
and contemplated by Kellogg over the course of forty years at
the National Museum. It's a rare day that I don't pause at his
contorted penmanship in India ink on brittle paper in a specimen drawer. His legacy, however, also includes the fact that he
was instrumental in founding the International Whaling Commission in 1946 and served as a kind of science diplomat representing the United States at the IWC until shortly before his
death in 1969.

Kellogg was well aware that some whale species, such as
North Atlantic right whales, had been devastated by Yankee

whaling less than a century before, and that other species, such
as gray whales, were on the edge of extinction in the early twentieth century. He successfully marshaled his colleagues to enact
the first multinational prohibitions for any further hunting of
right whales and gray whales in 1937 (with the organization
that presaged the IWC), but the scale of what Kellogg faced in
the post–World War II years of whaling far outstripped anything that happened in the nineteenth century.

In theory, the IWC was meant to regulate the killing of
whales; in practice, the IWC functioned more like an international hunting club. Nations managed whales as resources, the
way many fisheries do today, and for most member nations the
influence of the whaling industry outweighed any scientific interest in whales. To a degree, ignorance served their aims: if no
one really knew how many whales there were in the oceans,
there was no reason to abate the hunts. Kellogg's opposition
was paid little heed. By the time of his death, well over two million large whales had been killed; there may have been only a
few thousand blue whales left on the planet, less than 1 percent
of their population at the outset of whaling. The scale of this
loss of biomass in the oceans has no historical precedent. We
live in a world with far fewer whales than our grandparents, and
certainly than the world of our great-grandparents. The ecological consequences of this scarcity are still poorly understood.

Kellogg's role in the failures of the IWC during the midtwentieth century—especially as a paleontologist-turned-diplomat—
confounds and frustrates me. His portraits depict him true to
his bureaucratic type: posed at a large desk, with a specimen in

hand, eyes glaring from a dour face. Were the endless, quiet hours of committee work and travel a matter of pride in the service of scientific diplomacy? Or did the shortcomings of diplomacy eat away at him? We know little about any of his personal thoughts about whaling. His written reflections are dry and devoid of personal asides, sadly barren of the colorful language he used in the workplace, which dominates the memory of the handful of people I know who talked to him. While some superficial details about Kellogg appeal to me, they still don't help me understand him. Nor help to answer the questions that I want to know more than the others: What would I have done in his shoes? Was there anything to change about the fate of whales on Earth had I been in his place, at his time?

I ask myself that whenever I pass by the bones of USNM 268731 in the warehouses. The jaws of this massive, unparalleled creature came to rest here thanks to rapacious greed that killed over 99 percent of all other whales like it. USNM 268731 belonged to a ninety-two-foot-long female blue whale, harpooned off east Antarctica in 1939 by the whaling ship *Ulysses*. The vessel was a factory ship over five hundred feet long that Norwegian whalers operated over several years with observers from the U.S. Coast Guard, logging thirty thousand cruise miles. One sharp-eyed inspector corresponded extensively with Kellogg about the specimen collection that the ship amassed. He, or someone like him, must have taken note of USNM 268731's spectacular size, though we have no mention of these particular jawbones in anyone's notes. How exactly USNM 268731's jaws entered the collections—by boat, by crane, by truck—seems to be lost to history.

The best stories of scientific discovery are, at their heart, stories about people. There's no doubt that the facts of science are exacting and objective. But the narrative of how we figure out true things about the world is not necessarily clean and tidy. That's because scientists are real people too, with inner lives that sometimes bear on their work. Scientific discoveries happen in a social context, and they can be as much a matter of happenstance as the friendships we form.

Jeremy Goldbogen is one such friend who has profoundly influenced my life and my science. Jeremy is quiet and contemplative in the moments when I am loud and rash; I relish making him snicker over a profane joke. Although we have different skill sets—he's now a leading researcher in the discipline of biomechanics, or the physics of how organisms function—our careers have been closely braided, with lab work and fieldwork around the world tagging, dissecting, and digging whales. Friendships have origin stories too, and our shared quest to understand how whales became giants of the sea started on a walk in San Diego.

On a break from one of my first field seasons at Sharktooth Hill, I met Jeremy through a mutual friend in San Diego. Like me, Jeremy was wrestling with the insecurity of not knowing exactly what to study, but his problems were far away from bones. He had recently received a set of data from one of the first generation of suction-cupped tags deployed on rorquals just off the coast nearby—his colleagues had hoped to collect whales singing but instead found them feeding. At the time, no one had recognized that the tags recorded a trove of biomechanical data that could be inferred from changes in the speed of these feeding whales, but Jeremy would be among the first to figure that out.

One day on a walk to grab tacos for lunch, Jeremy asked me how hard it is to measure rorqual jaws. I replied that besides finding a long enough tape measure, it was mostly a matter of legwork and a strong back. But Jeremy was thinking more broadly about measuring energy; when these massive animals lunge with mouths agape, water loaded with prey pours in, generating drag at a major energetic cost. Because jawbones delimit the size of whale mouths, Jeremy figured we could theoretically use bone measurements to put real numbers on how much water these whales engulfed during a single lunge. Those data, combined with tag data from their feeding bouts underwater, would give us key information about the energetic trade-offs involved with lunge feeding. Within a few months, we pooled together some small graduate student travel grants and headed to the Smithsonian—a place that could help provide exactly that kind of information.

My quip to Jeremy aside, measuring whale bones is hard work. I envy my colleagues who work on land mammals, or even large reptiles—they don't need forestry calipers, transect tapes, ladders, and forklifts to take simple linear measurements of their study materials. Even elephant bones don't rival those of a large whale. Just measuring the length, width, height, or circumference of the bones and skulls of rorqual whales, even the smaller ones, requires thick foam blocks, moving straps, and all the coordination you'd need to move furniture. (Bring gloves; ditch the nice shoes.) It's minimally a two-person operation most of the time—something that Jeremy and I learned firsthand with our first visit to the Smithsonian.

Over the course of two weeks in the warehouses, we managed to measure every rorqual jawbone we could find, for a variety of rorqual species, small, medium, and large. One fact we determined out of data collection was that bigger jaws had less of a mechanical advantage than smaller ones. In other words, based on the simple lever mechanics of where the jaw muscles would pull on the jawbone, larger rorquals should spend more effort closing their jaws, a bit like lifting a bucket of water with a broom by holding it at its end versus somewhere down the middle. That made sense in terms of behavior: smaller rorquals, such as minkes, eat quicker prey and need to shut their mouths faster than blue whales, for example, which can afford to close their mouths more slowly around larger and slower-escaping swarms of krill. In life, it may have taken USNM 268731's ninety-two-foot-long owner as long as ten seconds to open and close its mouth around a volume of water equal to an entire lane of an Olympic-size swimming pool. The drop in mechanical advantage at this upper limit made us wonder about the limits to living life as a rorqual whale.

Jeremy turned to the *Discovery Reports*, to table after table of raw whaling station measurements, a grind of library work and hunt-and-peck accounting. He thumbed through every paper and every table in the whole series to find numbers directly comparable to those we collected in museum collections, such as jaw length, as well as others that could be measured only in the flesh, such as body length and the distance between the dorsal fin and tail fluke. (The most comprehensive series of measurements in the *Reports* were limited to fin whales.) Jeremy found that bigger fin whales could hold even more water in their mouths than you would expect for their body size. Assuming this pattern held for other rorqual species, it pointed to what was likely a benefit of being large as a lunge-feeding whale—the bigger the better. Commonsense logic added: to a point. There must be a trade-off for being big: benefits, but also limits.

Was something else at work here, something more than we could discern from dry bones and numbers? It was a bit like trying to know a bat from either its flight path or its skeleton without context—in either case, there is no substitute for seeing the skin of its wings and watching how it moves through the air. What we really wanted to know was how whales open their mouths underwater, nearly ninety degrees wide, to engulf a cloud of prey as large as their entire body in just a few seconds—and do it successfully many, many times every day. How do straplike jaw muscles control that movement? What happens, for instance, with the muscles that form the floor of the mouth? All of these questions required getting close to a whale in the flesh. We needed to see the whole organism to understand the muscles, nerves, and flesh that fire the amazing process of lunge feeding. Not just data points and bones.

10.

A DISCOVERY AT HVALFJÖRÐUR

I took a break from writing field notes to gaze out at the fjord. The long summer light of early afternoon spilled gold on the dark greens and gray browns of rolling farmland. Our field clothes, drying on a line, snapped in the wind that rushed down the hillside. I glanced over at Jeremy, immersed in his laptop, tugging the loose collar of his sweater over his mouth. "Hey," I interrupted, "want to go for a hike?" Jeremy waited a moment. "Sure," he said distractedly. Then he added in a low voice, glancing toward the kitchen, "Only if we bring beer."

There were no whales to dissect at the moment, a mile down the road at the whaling station. The catch boats were far offshore, and I knew that we both could use a break from data inventory and manuscript writing. We slipped on our boots and unlatched the gate. Walking by the clothesline brought the deep aroma of whale oil, even after several washes, despite the sunlight and brisk air. We started uphill against the ever-present wind, climbing over fractured piles of purple-brown basalt left by the lava flows that had created the bedrock of Iceland more than fifteen million years ago.

Less than an hour later, sweating as the North Atlantic wind raced over us, we sat down at the cliff edge overlooking the entire fjord. We shared the vista with fulmars, transoceanic birds that dutifully dive-bombed us to keep us away from their nests. Periodically one would catch a steady wind and soar eye to eye with us, over a cliff drop of hundreds of feet, the perfect profile of its wings keeping it effortlessly aloft. "Control surfaces, right?" I said to Jeremy. Animals moving through fluid, be it air or water, control their path using the same physics that explain how a plane uses its wing flaps. "Yup," he replied as he sipped the foam off a well-shaken beer. "Same as the flippers that humpback whales use to turn on a dime underwater." Fulmar wings and humpback flippers owe their similar functions to the power of natural selection, producing the same solution from different starting points—in this case, shaping wings and flippers out of ancestral forelimbs in seabirds and whales for moving in vastly different worlds.

From atop the mesa, we could see the whaling station ensconced at the end of Hvalfjörður, the long fjord that opened to the North Atlantic. "Hvalfjörður" literally translates from Icelandic as "whale fjord," though locals assured me that the fjord's name is centuries old and likely just for an errant whale passing by or washed ashore. The station had been constructed following World War II; during the war, the fjord's narrow entrance and deep coves were ideal shelter for convoys of Allied merchant vessels and warships against marauding German U-boats. I wasn't quite sure how big the whaling boats would look from this elevation, and the choppy waves were no decent guide for size either. Jeremy finally spotted the catch boat, cutting a tiny profile against the glare, its prominent wake and quick return signaling a successful hunt. We quickly finished our beers and scampered down the slope. Back at the cottage we pulled our dirty shirts and pants off the clothesline, and got in the car to head down the highway toward the station.

Whales had brought us to Iceland, but there was also the happy accident of sharing the same adviser, Bob Shadwick—Jeremy's main adviser for his doctoral work, and my postdoctoral adviser. As a comparative physiologist, Bob has the perspective of an engineer looking at the biological world. His expertise lies in understanding the inner workings of animals—the machinelike processes, such as heart beating, lung expansion, and the spring of leg muscles, that let animals live, breathe, and move. The three of us had already published a paper together before I finished at Berkeley, connecting our work on whale jawbones with tag data. We modeled the engulfment volume of one species (fin whales), using jaw size to quantify the energetic cost of a single lunge. We then calculated the implications over the duration of tag data, showing the incredibly high energetic costs of lunging, surfacing to breathe, and then lunging again. I was eager to expand on this work by connecting jaw and tag data for other rorqual species, and also to look at fossil jawbone data, which led me to a postdoctoral fellowship in Bob's laboratory in Vancouver.

Time with Bob and Jeremy meant stepping outside my comfort zone of fossils and sedimentary rocks to spend my days talking about blood flow and muscle action. Bob has a deep background in whales: in the 1980s a colleague brokered an opportunity for Bob and a team of scientists to study the enormous hearts left over from fin whale carcass rendering at Hvalfjörður. Whale hearts are built like all mammalian hearts—four chambers, split into a right and left set, with smaller atria and larger ventricles—but they dwarf all others in size. Imagine, for the largest of whales, a heart the size of a farm tractor tire. Your whole body could fit inside one of these organs.

Whales also rank on the high end of mammals in terms of blood volume to body size, and that volume is astounding, amounting to thousands of liters of blood. In any mammal, the muscle walls of the heart need to push that blood volume throughout the entire circulation system of the body—and send it all the way back to the heart on the return. Consequently, the blood pressure in the aorta (the first main pipe off the heart) is the greatest of any place in the whole circulation system. This pressure wave, however, drops off abruptly as the beat ends and then restarts with every cycle of the heart pumping.

The aorta has to be flexible enough to accommodate the sudden and repeated changes in pressure without blowing out. As mammals increase in size, both the thickness of the aorta walls and the aorta's diameter increase; the ratio of wall thickness to diameter holds steady across all mammals from mice to elephants, and whales follow this pattern too. In most mammals, the aortic arch has elastic properties thanks to a protein called elastin. What Bob and his colleagues discovered, however, was that there's more than just elastin in the aortas of gigantic whales: there is also a unique lattice of collagen sheets that provide extra, microscopic anatomical flexibility for the aorta to successfully withstand blood pressure straight from the heart.

For every special story we have about the anatomy of a whale, there are dozens more mysteries waiting to be solved. No one has ever recorded the beating heart of a wild baleen whale. We don't know how fast or slow the heart of a rorqual whale, such as the blue whale, races when diving, feeding, or even sleeping. These basic mysteries are as true for whales' hearts as they are for any other organ in their bodies—and with whales, it's the consequences of gigantic size on their performance that makes them so worthy of scientific investigation. Many stories still wait to be told.

Years later Bob had managed to secure an invitation for his whole lab to work at the whaling station in Hvalfjörður during the summer whaling season. This was our chance to work with whale jaws in the way Bob had with hearts, and to look for answers to a similar question: How do these parts work, biome-chanically? Jeremy and I spent months carefully researching, planning, and digging into the questions we had about the anatomy of these gigantic whales. Initially we wanted to understand the precise motions of feeding, from mouth opening to closure, including how muscles of the head connected and wrapped the jaws; the flexible yet locking jaw joint that seemed to snap open to ninety degrees but never farther; and the exact layering of muscles, beneath the fat grooves outside, permitting a parachute-like expansion of the throat pouch. Every one of these issues required studying, manipulating, and cutting away at a bus-size animal in a controlled way. Those aims were simply not possi-ble with carcasses washed ashore, even under the best circum-stances.

Working at Hvalfjörður meant finding myself at that same uncomfortable intersection between scientific study and commercial whaling as some of my forerunners. It wasn't without pause that I undertook the trip. Whaling in Iceland is commercial, and the whaling company at Hvalfjörður targeted fin whales, which are listed as endangered species—a legacy from twentieth-century whaling, especially in the Southern Ocean. In the North Atlantic, however, the population numbers roughly 50,000 individuals, meaning that Iceland's hunt quota of a maximum of 125 animals was not going to severely impact this population. (As many fin whales are likely killed every year by net entanglement and ship strikes.) Once stripped of their meat, the carcass remains were not valuable to the whalers but could be a gold mine for us. We wanted to intervene before organs and bones were ground up into bonemeal for fertilizer. Moreover, we knew that Iceland's renewed interest in whaling might be short-lived: how long would Iceland continue whaling if the commercial basis was a function of market prices for whale meat? Lastly, it was clear that these whales were going to be killed whether we were there to study their carcasses or not. I thought we owed it to ourselves, in some way, to at least be present to make the most out of the situation. Despite my misgivings, if I wanted to go I needed to get there soon.

We planned as though this trip would be our only shot, and as if we might get to dissect only a single whale. With others in Bob's laboratory, we packed our crates full of steel-toed gum boots, dozens of sample bags, gloves, and long knives. We felt prepared for measurements, at least, using the same surveyor's measuring tape and forestry calipers that we used on jawbones at museum collections. More critically, we planned on documenting our every move, which meant bringing cameras, tripods, and lots of backup batteries. Mostly Bob recommended bringing plenty of old clothes to throw away at the end. "You will not want to bring anything back, trust me."

At the station, we donned our bright orange coveralls and steel-toed, spiked boots in the locker room. We had beaten the first carcass to the deck, but only by a few minutes. There's nothing quite like the initial walk out on the deck of the whaling station, easily mistaken for a vision of hell. Billows of steam erupted from geothermal pipes that lined the walls of the station's open top deck, supplying water to continually rinse the platform and steam to power much of the machinery on the flensing platform and below it, including the enormous boilers. The pipe works derived their heat from the same geologic processes that built the basalt cliffs and fueled Iceland's active volcanoes—geothermal cracks that ultimately tapped the hot blister of tectonic plates, miles deep, that continue to form the island of Iceland.

The steel spikes on my forestry boots scratched and clicked on the wet asphalt, which had flumes of krill and whale blood running over it, along with the occasional chunk of unidentified gristle, all moving downstream to the slipway and out to the fjord. A steam-driven, twelve-foot saw blade waited to section parts of the carcass into slabs of muscle and bone, like the world's largest deli slicer. Winches clanked slowly as the inch-thick cables grew taut to pull each fin whale carcass, some forty to sixty tons in weight, up the slipway from the small dock.

On the deck a crew of over a dozen stood at the ready. Many Icelanders at the station represented two generations, a father and son, or an uncle and nephew. The directors of the show—the lead flensers, older men—issued sharp, quick whistles as the hulking seventy-foot-long whale carcass made its way to the platform, where several tons of deep red and purple meat, still steaming, would be removed and flash frozen in under two hours.

We waited patiently on the sidelines to sample and dissect. We'd been offered safety pointers: there was ample opportunity to lose a finger, receive an ugly gash from an errant flensing stroke, or fall through one of the openings into the boiler pit beneath the platform, its massive steel maw grinding bone and blubber into a hot mash. Bob, who is a great conversationalist, enjoyed making dissection requests of the flensing team: "Do you mind if we grab that artery after you're done?" Or "We really need to take a picture of that flap of muscle before you cleave it off." On days without a breeze, the thick aroma from the boiler pits engulfed the entire platform— imagine literally tons of cat food left in the hot sun—and never left our clothes. Or our nostrils.

Walking around the station at Hvalfjörður, I couldn't help but feel like I was in a time warp, listening to the hiss from car-size steam-driven winches and watching men in coveralls and spiked boots walk along the back of a landed fin whale. It could very well have been a scene from the whaling stations at South Georgia, or anywhere else, brought back to life. Even the precise, stepwise sequence of rendering an entire whale was a method replicated in its entirety from those whaling days: with each animal resting on its left side, they sliced away the throat pouch first, then decoupled the jaws, then, working toward the tail, pulled the organs out of the chest and abdominal cavities before dissociating the head, all the while carefully selecting deep purple meat from the back and tail to send down chutes for immediate freezing. Whaling technology today remains mostly unchanged from the midtwentieth century, but in Iceland it operates at a far smaller scale in terms of the number of whales killed.

After several weeks at Hvalfjörður, we had handled, measured, and dissected the parts of over two dozen fin whales. The chance for more than one clean dissection was more than we had hoped; dozens gave us the opportunity to understand the variation in anatomical systems. Other members of Bob's laboratory pursued their own projects on fluke and flipper measurements or parts of the vascular system while Jeremy and I focused on the heads. We had learned the rhythms of working when the whales arrived at the slipway and resting when the catch boats were out to sea. Sometimes the whales arrived in the middle of the very short Icelandic summer night, and we worked in a strange twilight; sometimes we stayed on our feet dissecting and wrangling blocks of tissue through two, three, or, on an unusual day, four whales. Coffee helped. After our first season in Iceland, we had already seen more rorqual whale anatomy than many of our colleagues had seen in their entire lifetimes.

Jeremy and I treated each opportunity like it was our last and jumped into the wicked theater of flensing knives and steel cables to take photographs of specific muscles and their placement, and to collect samples. We focused on specific questions with each specimen, looking only at the corrugated throat pouch in one instance, the roof of the mouth in another. Every whale carcass brought up the slipway in Iceland provided another opportunity for a new discovery; there is *nothing* boring about seventy feet of whale. Large whales aren't like smaller dolphins, where you have the ability to unzip their carcass in a controlled fashion, in a laboratory, with tools that don't require heavy machine operation. Nearly every time we dissected a fin whale at Hvalfjörður, we had a chance to answer some basic questions about how whales at these scales worked, because comparative biologists like us had never really had the time or opportunity to do so.

Much of what we handled and observed had not been mentioned in published literature. The best monographic treatments, over a century old, were only rough guides for structures that we saw with every whale rendered before us. Every observation of muscles inserting on bones, or nerves spread across sheets of tissue underlying blubber, set us clicking away on our cameras and writing in our notebooks, fingerprinted with oil. Each page of sketches, measurements, and remarks pulled back a little bit more of the veil of mystery around the anatomy of these animals.

Putting aside jaws and heads for a moment, we could look basically anywhere inside the chest cavity of a large fin whale and quite literally put our finger on a basic biological question that would require a dissertation to answer. Inside the rib cage, whale lungs line the top, like a bubble trapped at the top of a bottle. How much air does it take to inflate a fin whale lung? How quickly does it inflate? The answers to those questions have implications for how long whales could dive on a breath of air, which in turn would affect how long they could feed.

Or take the heart—why does blood flowing to the brain not pass through the carotid artery, as in every other mammal, but instead go through the rete mirabile, a noodlelike network lodged in the neck vertebrae? Why is a whale's diaphragm—the sheet of muscle separating the chest cavity from the stomach one—oriented in a diagonal way across the body?

Looking elsewhere, we would cut cross sections of flippers or tail flukes and find arteries enclosed in a rosette of veins—the classic sign of a countercurrent exchange system. This biological device is seen widely among animals, especially oceanic ones, including penguins, billfishes, and tuna, along with whales. It is essentially represented by a large central artery with smaller veins surrounding it, which allows returning cold blood, often extremely chilled by the surrounding temperatures at the extremities during a deep dive (or even in polar waters), to be heated by the outgoing arterial blood from the heart. How dense were these arrangements in any one part of the fin whale's body? How did they vary in individuals, or among rorqual species of different sizes? Did bigger whales have more rosettes for more body mass? No one knows.

Our main enemy in all of these circumstances was time: we had only a few hours to investigate a given part of the body before all of the whale parts went into the boiler pits. We needed to make decisions in the moment about which part of the body to pursue and what to sample—an engrossing challenge when hiking up mounds of rolling flesh to figure out the anatomy spilled out under your feet.

Toward the end of the first season, at the end of a day with two carcasses in a row, Jeremy and I contemplated the lower jaws, which were typically severed from the skull in the first steps of the flensing process and yanked to the side of the platform by an enormous cable. With little meat and a lot of bone, the jaws held little value for the whaling company. But they were important for us because we finally were able to see the bones we had measured at the Smithsonian wrapped in flesh, including the muscles that anchored them to the skull and the gelatinous tissue that comprised their jaw joints.

Whale jaws are important in understanding how whales have gotten so big: their jaws control how much food they can eat, which is a key part of not just attaining large body size but also maintaining it. It's also worth pointing out that baleen whale jaws are a bit different from those of most mammals, which have L-shaped right and left halves connected by tough fiber or bone in a continuous, interlocked arcade. Toothed whales' jaws form a simple V (or occasionally a Y), connected at their tips. Where they connect to the skull, they are thin and hollowed out—partially empty to house large fat pads that connect the jaws to the ears, helping the whale hear high-frequency sound as it echolocates.

Baleen whales have similarly simple, loglike jaws, lacking teeth. Where the right and left halves meet at the tip of the chin, there is only a clean surface and a small cleft of bone—a free range of movement at the tip that gives baleen whales' jaws flexibility, almost like snakes, in their ability to open and close their mouths. Feeding, not echolocation, is the primary job of the jaws in baleen whales. Where the jaws are anchored at the skull, large ball-like ends are wrapped in a mass of fibers, likely to accommodate an enormous amount of strain and torque. In rorqual whales this mass of fibers allows the jaws to open and close rapidly during the course of a single lunge. Jeremy and I figured that these jaw joints, which were hardly described in any kind of detail in the anatomical literature, were probably well worth investigating. We hoped to come up with some way to sample them, test their elasticity, or discern their microscopic structure on a histological slide back in Bob's laboratory.

At the end of a long morning at the station, in that sweet spot between exhaustion and boredom, we discovered something interesting—and entirely new. We were looking at an intact lower jaw—an enormous, silvery V-shaped structure with skin all the way around except for the knobby back ends, which were wrapped in expansive, white, fibrous tissue where they had been hacked out of the skull. Frustrated and a bit dulled by the fruitlessness of our effort to come up with a system to consistently measure gelatinous, floppy tissue, we opted for the mature solution and abandoned our efforts, deciding to hack away elsewhere on the fifteen-foot-long jaws.

Jeremy reminded me that at the tip of the jaws, where the right and left jawbones meet, at the crotch of the V, was supposedly a clean synovial joint, much like our own hips or shoulders—cartilage with a clean indentation for the smooth movement of two bones against each other. We recruited a few Icelanders to help us out, and they pulled the thousand-pound jaws in opposite directions as we made a clean slice in the middle. Suddenly, as the knives sank into the block of tissue, what we saw didn't make any sense: a tangle of fingerlike projections, pearly in color, spilling out of a cavity of some kind. There was not much blood, but instead gel-like goo that poured out of a core, surrounded by a fibrous wrapping that begged investigation.

"Whoa, what *is* that?" Jeremy exclaimed.

"I don't know, but it sure doesn't look like a synovial joint," I replied. Part of me wanted to ignore it, in a moment of befuddled disgust, but I knew better. Curiosity won out. "Let's go get Bob. This doesn't look right at all." I suddenly realized I had never considered what the inside of a whale chin really looked like because no one had ever mentioned it much in the literature. So goes discovery: unless you know your own ignorance, you won't recognize when something truly novel is right before your eyes.

The first thing we wondered was whether it was unique to the individual whale (maybe a pathology), so we wandered over to the jaws of a second whale, freshly removed from the current carcass on the platform. As we pulled a razor-sharp flensing knife across the jaw tip, yet another tangle of fingerlike projections popped out. I looked at Jeremy, who looked back at me

with eyebrows raised. "You'd think someone would have described alien goo leaking out of the jaws of these animals, after
a few hundred years of whaling."

We scoured the anatomical reprints we had on hand for any
kind of description that matched what we had seen. With one
or two exceptions, it seemed that no one thought the jaw tips
were interesting, and the exceptions didn't provide great photos
or extensive discussion. With every new whale that landed on
the flensing platform, we teased at this enigma a bit more—
quite literally with forceps and sample vials—and we took
plenty of photos of each jaw tip, split down the middle, with
scale bars. Eventually we came to realize that we lost important
information with every knife cut—it was a gelatinous structure
sitting in a cavity bounded by a hard exterior, like a jelly-filled
candy. We needed more controlled circumstances for our dissection than the asphalt and bustle of the flensing platform. Bob
had a brilliantly simple answer: why not cut off a whole jaw tip
and ship it back to Vancouver for study in the lab?

We settled on one of the largest individuals we had yet seen.
I recruited some Icelanders, who were nonplussed by the re-
quest, to use the twelve-foot-long saw blade to cleave off the
whale's chin, separating it from the jawbones and the throat
pouch. It fell to the platform, a pyramidal hundred-pound prow
of whale flesh. Inside, I hoped, was the intact geometry of whale
anatomy that we didn't understand but that was in all likelihood completely new to science.

11.

PHYSICS AND FLENSING KNIVES

Imagine, for a moment, trying to feed like a rorqual whale. Rising up from the deep, you fluke at full speed to confront a massive swarm of zooplankton. Imagine dropping your massive jaws, opening wide, as if preparing to take a bite of an apple. Now try to imagine an incoming rush of water pushing your tongue down into a pouch between your throat and the skin around your neck. As your tongue slides down, this pouch expands in a giant bolus that reaches all the way to your belly button. Then it balloons outward, holding a volume equal to that of your own body, like a boa constrictor after swallowing a deer. Eventually, over the course of a few minutes, the muscles of your body and neck contract around that bolus to strain any prey inside through your mouth and swallow your meal. That's exactly what happens to rorquals when they take a gulp during a lunge—and they do it dozens of times each day of their lives to feed.

Feeding is everything where most animals are concerned. It's surprising how little we actually knew, until very recently, about how it works when it comes to rorquals. Tag data have allowed

inferences about their
behavior that we cannot see below
the surface: how deep they dive for food (as
much as a thousand feet), how long each foraging
bout lasts (between five and fifteen minutes), how
fast they attack their prey (topping out at about ten
miles per hour, a speedy six-minute mile for us), and how
long it takes to strain that gulp of water and prey with
their baleen (a few minutes). Tags have also clued us in to
underwater acrobatics that rorquals use to capture prey, including 360-degree barrel rolls by blue whales. But tags can't really
tell you about the underlying anatomy that underpins all of this
behavior.

Many whales have a pair of pleats on their throats, but rorquals uniquely possess dozens of throat grooves, technically
called ventral groove blubber, which run from their chin to
their belly button and form the external surface of the throat
pouch. When rorqual whales lunge, the ventral groove blubber
allows the throat pouch to expand outward, like the pleats in
an accordion. The grooves have thick and resistant primary
ridges separated by soft tissue; as a result, the outstretched
throat pouch has a stepped, corrugated texture. During a lunge,
the pouch's expansion is sudden and dramatic in scope; it
ripples a bit, almost like a parachute opening. My colleague

Jeremy wrestled with a tricky question that arose from the comparison: did rorqual throat pouches open passively, snapping open in response to the sudden rush of water, or could a rorqual control the intake actively, pushing open the muscles in its throat, like an egg-eating snake?

Jeremy connected with Jean Potvin, a particle physicist turned parachute experimentalist, for help. Among other work, Jean has tested very big parachutes for the military and for the skydiving industry. Jean also takes his academic interests seriously in an applied way: he tests many of his own parachute designs, whether it means sitting in the open end of a cargo plane or completing one of the 2,600 parachute jumps he's made himself.

Working together, Jean, Jeremy, and Bob applied the mathematics of parachute physics to the rorqual throat pouch. One of the more complicated aspects of the comparison involved the fact that over the course of a single lunge, the mouth lets different amounts of water into the throat pouch because the amount of the mouth exposed to flow changes as the jaws open and close. The data from our museum jawbone measurements, along with the *Discovery Reports*, also proved remarkably useful here as parameters for jaw length, mouth width, and throat pouch length, which Jean needed for computing the physics of throat pouch expansion. The measurements were also helpful because they spanned a range of rorqual sizes, from minke to blue whales. Jean could then calculate, for example, how well the parachute-like throat pouches worked at different sizes, and whether there was a limit to how large (or small) they could be.

The verdict on passive versus active throat pouch expansion? It most definitely could not be passive inflation. Drag is the primary physical force involved with parachutes and throat pouches—it's a type of friction that acts in opposition to the movement of a body through any kind of fluid, whether it's air or water. Drag on airplanes or dolphins is relatively low because their form is streamlined, but the drag created by essentially the inverse structure—a cup or a parachute—is greater than that of a flat panel moving through water. We already knew that lunge feeding was energetically costly because of drag, but Jean's calculations showed that the scale and rate of throat pouch expansion would result in a catastrophic blowout if only passive forces were involved. In other words, whales have to employ active resistance against the oncoming rush of water, full of prey.

It made sense that the muscles lining the throat, underneath the corrugated exterior, provided that kind of active resistance. In life, when not feeding, whales hold the throat pouch trimmed to their streamlined form; in death, the throat pouch is slack, floppy, and unyielding to manipulation without meat hooks and flensing knives. In Iceland we spent a lot of time studying the throat pouch. It was, conveniently for us, one of the less important structures to whalers. We had to plead with the flensing team for time on the pouch because it was a gigantic obstruction covering their access to the rest of the carcass; they usually removed it quickly, as they did with the jaws.

Beneath the corrugated exterior of the ventral groove blubber are three layers of muscles, what would be called platysma in most mammals—a thin sheet of muscle that makes the whole body of, say, a horse quiver in the cold air. In humans the platysma runs only from our jawline to our neckline and can easily be made taut: our shaving muscle. In rorquals the platysma's three layers can contract side to side or front to back—a key to maintaining and controlling the overall shape of the throat pouch, which, when inflated, can hold a volume of water equal to that of the entire whale.

In sorting through the muscles beneath the ventral groove blubber, we noted another structure that had been described in the literature before but not explained: right and left branches of a thick, hard Y-shaped structure embedded in the blubber layer. It was subtle but real, and we didn't know what to make of it. The right and left branches ran parallel right below the jawline, while the stem, connected at the midline, ended below the chin—an anatomical coincidence that piqued our interest, especially after having seen the strange structure inside the chin. Were they connected somehow?

Every time we tugged on one anatomical problem in Iceland, we found several more. We were left with a mounting pile of mysteries about the anatomy underpinning rorqual feeding, and I wasn't sure about the way forward. What's more, we were examining only one species. We needed to replicate our findings, ideally with evidence from another species of rorqual whale. I wasn't entirely sure how we would do it. To understand how nerves played a role in lunge feeding more generally, the ideal would be to collect the freshest tissue possible, like that taken straight from a harpooned whale, on a catch boat.

In the middle of the North Atlantic, far from sight of land, Jeremy and I huddled together on the top deck of the *Hrafnreyður*, a minke whaler. I steadied my forearm and set my scalpel into a jagged chunk of whale tissue taken from the throat pouch. Smaller muscle bundles twitched as nerves fired from still-living cells; the minke whale had been killed by a harpoon less than a half hour earlier. I focused on the facts at hand: twitching muscles were a good sign that we would be able to collect high-quality microscopic information. I placed the slice of muscle onto a metal plate cooled by dry ice in a Styrofoam box, which Jeremy quickly closed before holding out a formalin vial to collect another specimen to fix. Just then, the boat heaved to change course. We tried to hold steady in the churning waves as the diesel engine whined. The skipper leaned out his window to yell at the spotter dozens of feet above us in the crow's nest. At the prow, the skipper's seventy-six-year-old father stood rigid, legs braced apart, bundled in his dingy survival suit, his gaze fixed on the lapping, mind-numbing waves. Another minke had been spotted.

Being on board this commercial whaling ship was a once-in-a-lifetime opportunity to collect material from as fresh a specimen as possible. I actually had two aims: collect fresh nervous tissue from the throat pouch, and see if the sensory structure Jeremy and I had found in fin whales showed up in other rorquals. Nerves decay notoriously quickly and aren't useful for making stained microscope slides unless they're fixed immediately after death. And we knew that if we could find the strange chin structure in minke whales, there was a much stronger case for inferring that all rorquals probably possessed it because of a common evolutionary origin. In other words, was the structure just a novelty for one species, or was it something found in all rorquals, perhaps contributing to the broad evolutionary success of the whole group, over millions of years? In that moment, nobody knew because no one had ever seized on the gooey, messy structure the way our research team had.

The smorgasbord of blades, vials, and boxes set out on the whaling ship's deck represented the culmination of years of planning. But as I looked out at the flat, gunmetal skies, I thought about how our improvised, open-air lab bench was in the service of work similar to what so many other scientists had done aboard a whaling boat decades ago. I was comforted to have Jeremy along, a close colleague and friend to share the experience.

Jeremy reached to grab a thick marker to label our vials. "We really need to seal these vials with—" *POW*. A chest-pounding clap of thunder hit us. I tucked my head into my shoulder; Jeremy ducked momentarily while still managing to hold his vial upright. After a few beats, we both looked around, realizing that another harpoon had been shot. A black cloud of smoke descended on us from the bow, wrapping us briefly in its acrid smell. The gunner looked back at us, his face covered in soot, as he whistled toward the other crew, pointing to the successful strike.

A harpoon grenade exploding in its chest cavity killed the minke instantly. The thirty-foot-long carcass hardly fit on the back deck of the *Hrafnreyður*; the crew quickly set to work to remove long, dark backstrap muscles prized for meat. It was a bit like watching a miniature version of what we had seen happen with larger fin whales at Hvalfjörður. Little about the minke whale, other than its choice cuts, held any interest for the crew.

I seized the moment and asked if one of them could collect the chin for us. I cradled the chunk of flesh, the size of a shoe box, and brought it back to the main deck. "All right, here we go," I said to Jeremy as I pulled the knife across the *V*-shaped intersection at the tip of the jaw. The cut revealed a distinct cavity, full of pearly papillae. We smiled at each other in wild recognition—the structure was here. It wasn't unique to fin whales; it was probably a feature in all living species of rorquals. We were exhilarated, even though we would later discover that the nervous tissue from the throat pouch had decayed too much to be useful for the microscope slide work. Nevertheless, we had new information about what was going on inside one of the largest animals on the planet—and now we could try to figure out what it meant about how they live.

When we returned from Iceland, we began sorting through hundreds of pounds of frozen tissue in Bob's lab, all mediated by the appropriate permits for transit. We conducted some practice runs thawing and studying less critical material first. But then it was time to examine our hundred-pound fin whale chin from Hvalfjörður. What was going on inside that gooey structure, especially the freakish fingerlike papillae—did they house nerves, along with what seemed to be blood vessels? Why would there need to be blood flow inside a dense and rigid part of the chin? We needed a closer look.

We wrapped the chin in several rolls of plastic wrap, bubble wrap, and large plastic bags, with rolls of paper towels tossed in to soak up fluids. We then hauled it next door to a giant CT scanner that had been custom built by a forestry company in Vancouver to scan entire tree logs from the Pacific coast. We knew that the chin would begin to thaw while it was being scanned, but because the soft parts were embedded in a stiff outer layer, we didn't expect the structure to deform or decay too much. In subsequent MRI imagery, we saw the exact path of what looked like large vessels, from a canal in the bony jaw tips toward the soft cavity in the midline. The MRI and the CT imagery together gave us two excellent road maps for making our first cuts. It was a dissection no one had ever done before, and we wanted guidelines to the anatomy, where to begin, what to remove, and what might reveal the most telling information.

After a days-long dissection following our digital maps, we had all of the clues we needed to start solving the mystery of the alien goo trapped in the chin of a whale. First, we learned more about the structure lodged at the point in a rorqual's chin where their bony jaws meet—a structure with a soft core, loaded with fingerlike structures and encompassed by thick, hard connective tissue. This structure was not a joint, and somehow no one had remarked on it before—among all the hundreds of thousands of rorqual jaws that had likely been split apart by whalers. We asked colleagues for photographs of the structure from any other species of baleen whale, not just fin and minke whales; someone else had likely seen this structure during a necropsy at a whale stranding but probably hadn't thought much of it because of the decay. (The fact is that, most of the time, whale anatomy is either too decayed to be useful or inaccessibly lodged under a mountain of body parts.) Other rorqual species had it, while distantly related baleen whales, such as bowhead and right whales, did not—they had a few papillae but no distinct gelatinous cavity. The scorecard of which species had the structure and which didn't was strong evidence that the structure we had found in rorquals was a result of a common evolutionary origin.

Second, we knew from both the digital imaging with the CT and MRI scans and the wet dissections that the soft cavity was wired with nerves and blood vessels. When we looked at them on a microscope slide, we saw that the papillae were loaded with pressure sensors, also called proprioceptors. All mammals have them; the whiskers of a cat, for example, end in them. They look like tiny, coiled structures at the cellular level. Pressure sensors essentially tell your body where you are by detecting the body's position from movement and relaying that information to the nervous system. That kind of feature, carried inside a gooey cavity within the chins of rorquals, probably had something to do with telling a whale what was happening with its jaws.

And, just as a strange aside: the anatomical wiring of the soft cavity was asymmetrical. At the whaling station, we had realized that our clean orthogonal cuts through the chin, like slicing equal slices of bread from a loaf, were somehow always revealing bundles of nerves and blood vessels spilling out from one side or the other—never symmetrically on *both* the right and left sides. I realized that we had enough replicates with each individual whale to score this asymmetry, which was predominantly wired to the left in fin whales. (Even the two minke whales we dissected at sea were not uniform.) Asymmetry seems to happen in whales in funny ways—narwhal tusks are usually off to one side, and sometimes baleen whales feed exclusively on one side.

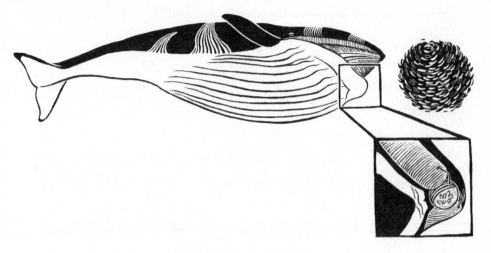

A rorqual's sensory organ

None of these examples, however, explains the asymmetry we observed—a mini mystery still waiting to be solved.

Eventually we had enough information to start calling the soft cavity structure a sensory organ. It certainly possessed a variety of different tissues, which was the first step in matching the criteria for an organ: we had nervous tissue, blood vessels, and clusters of papillae that floated inside a soft cavity pinched between the bony tips of gigantic jawbones. Now we had questions about its function.

The sensory organ is loaded with nerves, some of which enervate the tiny hairs on the chin. Once a rorqual finds a large swarm of prey, individual krill or fish on the periphery of this swarm brush by these hairs, tipping off the whale to open its mouth, in anticipation of hitting much greater densities. (Seals, like cats, have whiskers partly for sensing prey without sight.)

As the whale's jaws open, the soft part of the sensory organ compresses, and the nerves inside it pick up on that shearing, sending information to the brain. As water rapidly rushes into the mouth, the expansion of the throat pouch pushes on the thick, rigid fibrocartilage pads embedded in the ventral groove blubber directly under the chin. A common root to the pads (which form a Y with a right and left side) sits right below the sensory organ and likely provides information about just how full the throat pouch is and when the whale should begin to close its mouth.

In some ways, having an extra bit of nervous control makes sense when your jaws need to open and shut in a few seconds to make a meal—all the more so when those jaws, in some species, are the length of a living room. But what does having a sensory organ mean for how these whales got so big? The sensory organ coordinates the entire complex anatomical bizarreness of a lunge, allowing rorquals to pull off one of the largest biomechanical feats on the planet—every day. The structure is notably absent in baleen whales that aren't lunge feeders. Toothed whales in particular have nothing like it. Given that the whale family tree includes a few gigantic whales that don't have the sensory organ, such as right whales and sperm whales, it would seem that it isn't necessarily a prerequisite for evolving enormous body size. But it certainly doesn't seem to hurt; it's clear that the common ancestor of all of today's rorqual species had such a sensory organ, and we know that extinct ancestor wasn't particularly large for a whale—perhaps the size of a minke whale. When, then, did rorqual whales get so big? Answering that question, and the evolution of giants in general, requires looking outside squishy tissue, tags, and bones.

12.

THE LIMITS OF LIVING THINGS

We tend to think of gigantism as a feature of the past, almost as if the extinct ancestor of every species alive today was larger, gnarlier, armored, or equipped with saber teeth. We also gravitate to news stories about the fossil record that focus on superlatives—gigantic size being one of the perennial favorites. These tendencies of ours have a bit to do with the fact that iconic mammals like mammoths, cave bears, giant ground sloths, saber-toothed cats, and even the largest dinosaurs are no longer around—we don't see behemoths unless they're in the zoo or when they're in protected enclosures (or see their bones in a museum). But we're actually living in a time of giants right now. Blue whales, fin whales, right whales, and bowhead whales—all cetacean species that were targets of whaling at one point or another—are the largest animals, by weight, ever to have evolved. They just don't happen to live close to anywhere where most people would see them. Even the largest animals of all time are still elusive and rarely seen on an instrumented planet with several billion people.

Every lineage of gigantic organism has evolved from smaller relatives. There's no rule that says giants don't sometimes beget smaller descendants—island dwarfism, after all, sometimes happens in mammals—but whether it's dinosaurs, elephants, horses, or even rodents, a general trend toward gigantism holds. (It appears less common as a pattern in the invertebrate fossil record.) This feature in the history of life has been called Cope's rule, for nineteenth-century paleontologist Edward Drinker Cope. (It's unclear if Cope deserved having his name fixed to a rule after he first observed the tendency of lineages to get big over geologic time, but he named a few fossil whale species, so he's a sympathetic figure to me.) The gist of Cope's rule, as elaborated by his intellectual descendants, holds up as a testable model for a major mode of how evolution works on this planet. But merely saying animals get big over time isn't explaining much—it's the when and how that really lead us somewhere interesting.

It took dinosaurs tens of millions of years to achieve the leap in size from lapdog to elk, and then, after having been around for another fifty million years, they achieved the titanic size classes of tens of thousands of pounds. Land mammals, by contrast, evolved their largest body sizes relatively quickly, within ten million years; every time they dispersed to a new continent, they attained the same size class (around that of an elephant) in about the same amount of time. Aquatic mammals, such as whales and sea cows, show an inverse pattern—it has taken nearly their whole evolutionary histories (some fifty million years apiece) for the largest exemplars of whales and sea cows to evolve.

While we think of our own distant evolutionary forerunners, including *Australopithecus*, as being relatively small compared with us, all of the extinct relatives in our own family tree

were essentially within the same magnitude of body weight—
the same categorical bin—since our divergence with chimpanzees about 6.5 million years ago. At most, there's arguably a
onefold increase across the full range of fossil human ancestor
sizes. Elephants increased in body weight about a thousandfold
from the first elephant relatives that lived in Egypt (found in the
very same rock units that hold remnants of *Basilosaurus*) to
Siberian mammoths, which went extinct only at the last ice age.
Other land animals essentially undergo the same scale of transformation, whether rodents or sauropod dinosaurs. But whales
are clearly in a separate category with a ten-thousandfold increase in size since the time of *Pakicetus*.

When we want to understand life as a giant—a gigantic whale,
mammal, dinosaur, or anything else—it's useful to categorize
the consequences of size (that is, the advantages and drawbacks)
as seen either inside or outside the organism. One obvious re-
striction on internal functioning for giants on land has to do
with gravity, as it constrains bone growth, blood circulation,
breathing, and any aspect of reproduction. One obvious external limiting factor is sourcing food. After 100 million years,
sauropod dinosaurs seem to have encountered a genuine limit
to growth at around 110 feet in length. Somehow, it's hard to
imagine how trees could sustain the demands of a single herd of
gigantic sauropods, let alone multiple coexisting species.

One obvious advantage of large size is predator deterrence:
at sufficiently large sizes, whales will be too great a risk, or simply immune, to predators like killer whales or, in the distant

past, megatoothed sharks. Another advantage is physiological efficiency. Locomotion and migration become more efficient at larger body sizes—that is, more distance can be covered with proportionally less energy expended. Plenty of other ecological consequences of body size come into play with whole-organism considerations: the largest whales are so big and thick with blubber that their physiological challenge is shedding heat across their skin, instead of retaining it. For whales feeding and swimming in polar waters or diving to depths of a mile, energy-storing and heat-trapping blubber is a lifesaver; but in warmer latitudes or in shallow environments, it likely constrains much of their lives. Most whales don't live in any one place very long: large whales migrate huge distances to feed, which implies a careful strategy to maximize the benefit of blubber for storing energy over periods without food while minimizing the cost of overheating. The basic physics of heat transfer have implications for aquatic life: water conducts heat loss twenty-five times faster than air. This knowledge makes it possible to calculate a hard lower limit on just how small a whale can be while maintaining enough heat to live—fifteen pounds, a floor that the birth weights of even the smallest species of living cetaceans appear to obey.

It turns out that a lot about an animal's biology—how quickly its heart races, how many young it produces, how long it lives—can be predicted from its size alone, whether the animal is living or extinct, enormous or microscopic. Smaller mammals tend to have high metabolism, give birth to many young, and live short lives; large ones—including whales—tend to burn calories more slowly, reproduce less frequently, and live longer. The mathematics that describes how biology changes across these scales of size has been called allometry. Many of the predictive features of allometric equations tend to relate to general physical principles, like the rules of how heat dissipates or the ratio of an object's surface area to its volume.

Consider the fundamental constraint of the surface-area-to-volume ratio: no matter the shape of an object, its surface area always increases slower than its volume. One of the first biological implications of this relationship for any organism larger than an amoeba is that gas diffusion alone can't work fast enough to bring oxygen to all of the parts of the body, necessitating some kind of mechanism—a machine, such as lungs—to stay alive. Whales certainly have lungs, which means their answer to this latter problem is mostly in line with that of any other mammal, but their lungs are large enough and specialized enough that they present other quandaries. Whale lungs, as far as has been investigated in smaller species, have structural modifications that allow them not only to collapse quickly enough to avoid buoyancy problems and tissue rupture when they dive over a mile deep but also to reinflate rapidly at the surface afterward.

In line with breathing, another major challenge is holding enough oxygen to spend as much as two hours underwater for the deepest-diving whales. Whales overcome this challenge through a variety of anatomical and physiological solutions at different scales, such as high ratio of blood to body volume, high blood cell count, and blood cells with elevated hemoglobin concentration—that is, blood cells with oxygen-storing enhancements. (In another example of convergent evolution across very distantly related branches of the mammalian family tree, these remarkable blood traits evolved independently for deep-diving mammals such as whales and seals.)

On land, the force of gravity imposes another fundamental constraint. As organisms scale up, physics dictates what's possible for any kind of movement and function, be it blood flow, digestion, or locomotion. Sauropod dinosaurs, for example, had columnar limbs to support their massive weight, yet likely lightened the load by deploying an avian respiration system, which permeated their skeleton with air sacs, as in birds today. Whales obviously haven't had to deal with the force of gravity since they became fully aquatic, making them essentially weightless underwater. Instead, forces such as drag have shaped their bodies, especially when feeding.

Applying allometry to the study of whales—especially baleen whales, including the largest ones ever—is the key to understanding not just what it takes to be a giant but also the limits of living things.

When Jean Potvin used allometry to calculate drag on mathematical models of different-size rorquals, he found that beyond lengths of 110 feet, a blue whale would not be able to close its mouth fast enough around quickly escaping prey, nor make up for the huge costs of drag and energy lost from the act. In other words, the 109-foot lengths of the largest whales ever measured seem to correspond to a theoretical maximum for living life as a blue whale.

Jean's work explains a bit why we don't see, say, two-hundred- or three-hundred-foot-long blue whales in the oceans, but there are other possible reasons beyond the biomechanics of feeding. For example, large body size ought to confer better diving abilities, but rorquals do not dive anywhere close to the depths expected for their body size. In part it's because their prey are located farther up in the photic zone, but it also seems that the energetic costs of lunge feeding at those great size classes—the burden of dealing with ever-increasing volumes of engulfed water, as well as the necessity of breathing enough oxygen at the surface to make up for depleting all of the air while lunging—seem to impose hard limits despite the benefits of filter-feeding efficency at large size.

Take one step back to consider what we know about the factors promoting extreme size in whales over the course of their history. First, the anatomical innovation of baleen, which likely evolved around 30 to 25 million years ago, may have conferred important physiological efficiencies for its first adopters, but it wasn't a serious breakthrough in terms of body size—the first baleen-bearing whales weren't that much larger than their toothed predecessors. What about other innovations, especially anatomical ones such as ventral groove blubber or the sensory organ in the chin? All the anatomical gear that appears so critical to rorqual whales—a whole suite of specializations that range from new organs to bungee cord–like nerves in the throat pouch—likely had an evolutionary origin sometime during the late Miocene (around the time of Cerro Ballena). While they seem necessary in the sense that these features are prerequisites for being a blue whale (just like baleen), they did not automatically confer gigantism. Returning to the previous construct: If gigantism in whales and in other creatures isn't explained solely by factors inside the organism, what about *external* causes?

Typically, humpback whales migrate from the tropical latitudes of Hawaii to the panhandle of Alaska every year. They rest, mate, and give birth in the tropics during the winter, then navigate—by stars, by the Earth's magnetic field, by acoustic or visual recognition or a combination of those, we still don't really know—to the outer and inner coasts of the Alexander Archipelago, off Alaska. They arrive in the spring and gorge themselves on the herring runs. It's a long trip, but it seems to be worth it; today, no longer on the U.S. endangered species list, they reliably show up in large numbers.

One spring Ari and Jeremy invited me to join their team of researchers on a humpback-tagging expedition aboard the *Northern Song*. Humpbacks in Alaska have been known to hunt together using bubble nets—literally a curtain of air bubbles emitted at depth by one or more individual whales moving in a tight circle. As the bubbles rise, they corral a school of fish in a cylinder; humpbacks then gorge themselves by lunging upward inside the cylinder, about as close as they can get to feeding on fish in a barrel. These coordinated feeding groups of whales aren't particularly stable—they'll assemble and disassemble randomly—but the behavior is certainly learned and transmitted from humpback to humpback across ocean basins. Tagging was one way to capture some basic information about this amazing and mysterious behavior, which some scientists even describe as a kind of humpback culture.

Beyond a valuable opportunity for tagging, the expedition was also an opportunity to put my phone away, ignore e-mail, and spend the entire day talking, mostly about science. (Mostly.) The trip was the opportunity to push at the questions sitting on the edges of our disciplines, at the intersection of Ari's understanding of behavior and local ecology, Jeremy's grasp of physiology and biomechanics, and my background in paleontology and Earth history. The basic questions about how whales do what they do require all these fields of understanding, and I've always thought the best way to answer them was through this Venn diagram of disciplines and personalities.

This expedition had an additional perk for me: it was my turn to tag. Jeremy assembled and handed me a carbon-fiber pole equipped at the end with the latest tag, a piece of neon-pink plastic with cameras on the front and back, able to record everything going on in multiple directions at the same time—including other whales nearby. The pole felt heavy and ungainly. I felt like I was sweeping the air with a twenty-foot-long broom that held a five-pound weight suspended at the end. The next step was to somehow stand off the bow of a boat as an eighty-thousand-pound wild animal moved in the water a pole's length away, and affix a tiny piece of plastic to it. I was nervous, like a rookie before a game-time debut. Ari smiled and tried to reassure me with slow, deliberate instructions.

We didn't really need to do much chasing. The whales, having ended a feeding run, simply stalled and bobbed at the surface. Ari picked the nearest one and maneuvered the boat slowly. My muscles ached from staying poised and ready holding the pole. Then the gap started closing rapidly. I heard Ari say, "You're good," several times until I finally decided to go for it. I leaned out, aimed for a spot right behind the dorsal fin, and did my best to heave the pole upward to slap the tag on. My effort ended up being more a poke than a *thwack*. As a consequence, the tag didn't unlock properly from the pole tip and, reacting to the prod, the whale raised its fluke to dive, barely missing our boat. I winced as the pole creaked awkwardly, discovering that I had in fact bent it. "That's okay." Jeremy smiled. "I'll bill the Smithsonian," he teased.

We watched the whales swim off, until the closest one rolled to the surface. "Hey! Tag on!" I yelled. I caught a glimpse of the neon housing clinging to the whale's side before it sank back down into darkness. I slumped in the boat, exhausted and happy as Ari cheered. Jeremy snapped a rapid burst of photos as the humpbacks fluked up and dived deep.

Toward the end of the day, we sat atop the bridge of the *Northern Song*, binoculars and notebooks out. The peak of the herring run in southeast Alaska meant abundant food for predators of all stripes. The *Northern Song* cruised leisurely into Seymour Canal, a cove off the inner coast of the archipelago. We watched bald eagles by the dozens drop down from spruces to pluck herring thrashing at the surface. They wheeled in a wide spiral, the column giving a clue about what was happening below the water's surface. From the height of the bridge, we watched one, two, and then three humpbacks surge from depth and lunge

with mouths agape. It wasn't exactly coordinated bubble netting, more happenstance opportunism. As they closed their
mouths, their corrugated throat pouches sagged at the surface
until one by one the whales tightened their slack pouches and
returned to the deep. Looking around, we realized that we were
surrounded by dozens of humpback whales, far into the distance, all gorging on their own balls of herring.

It was an astonishing sight: whale snout after whale snout
breaking the surface, jaws lagging behind like a scoop, along
with trumpeting blows and flashes of scalloped flippers. I
couldn't count the number of blows firing off around us, all the
whales surfacing in the cove. "Pretty amazing," I said aloud, as
if I needed to; everyone else gazed on the scene quietly, some
with camera lenses pointed down. The scene felt almost primordial, ripped from a novel about other worlds; but in fact the
factors that let these whales enjoy their herring-run feast were
all very recent in the time line of life on our planet. And these
factors, which explain why we live in an age of giants, have to
do with glacier fields lining mountains, far off in the distance.

After sorting through all of the possibilities, we came to see a
clear shift in baleen whales' body size, toward gigantism, that
happened in the last 4.5 million years. Graham Slater conducted
different mathematical tests to see whether our data could be
explained by any other models of size evolution—an early burst
in large size, a bias against preserving large whales, or a simple
diffusion of this trait over the course of millions of years—and
found that the relatively recent jump toward very large body

size in baleen whales was probably real. Put another way, the dramatic increase in whale body size—*the* thing that makes whales iconic, in many ways—happened only in the last few moments of their evolutionary history.

This time frame in Earth history is important because it marks the start of profound changes to the oceans during the ice ages about 2.5 million years ago. We still live in this time of alternating glacial and interglacial periods, which seesaw in timing over the course of cycles determined by the periodic wobble of the Earth and variation in its orbit around the Sun. On land, the march and retreat of ice sheets to the latitudes of Seattle, Chicago, and New York City, has meant seasonal periods of melting and freezing that increased the sediment load for freshwater river systems emptying to the coast. Coupled with stronger wind patterns that have also powered upwelling, these changes to the oceans, only a few million years ago, set the scene for time-sensitive and dense aggregations of zooplankton to make coastal seas highly productive at certain times of the year (krill swarms reach their apogee during the summer months, regardless of the hemisphere). In other words, the iconic world of whales we see today off Alaska, Monterey Bay in California, or Stellwagen Bank along coastal Massachusetts happened only with the conditions enabled by recent ice ages. The ice ages made the distribution of oceangoing prey more highly concentrated in local regions over shorter amounts of time.

Given all the costs of living life as a rorqual whale, the largest among them, such as blue whales, need to make the greatest return on their feeding investment, which means heading for the highest densities of prey. Once there, the efficiencies of filter feeding with baleen, and the anatomical specializations that co-ordinate lunges, become critically important for maximizing the capture of prey that varies in availability. Being a rorqual is also just one way to take advantage of this high-density prey availability in the modern world; large size among other filter-feeding whales, such as right whales and bowheads, confers the same advantages.

Being big also means efficiencies in transit to the farthest of these swarms, fending off predators while en route to maximize their yield with massive lunges using increasingly large throat pouches (and mouths). The patchy availability of prey enabled by ice age seas provides an ecological explanation for why many different baleen whales all seemed to get extremely large within the same time frame of the past 4.5 million years—the last few moments of their over-50-million-year history.

The worlds that whales lived in made them big. But large body size, for land mammals in today's world, is correlated with a higher extinction risk. Extreme body size can be an overspecial-ization itself, requiring vast amounts of resources to sustain. As the largest creatures on the planet, whales live at the knife edge between perfect and perilous adaptation. Their large size can be a liability if the environment changes rapidly, a fact that makes this kind of understanding important, as massive changes to their world are happening, quickly, thanks to the behavior of our own species. How successfully whales and humans can share this evolutionary moment in the age of giants is a high-stakes story that's still being written.

PART III
FUTURE

13.

ARCTIC TIME MACHINES

In 1846 two tall ships, sails full of cold air, slipped through a passageway choked with black-stained ice floes. The HMS *Erebus* and HMS *Terror* were a year into an expedition led by Sir John Franklin in search of the Northwest Passage. Franklin's ships were marvels of the latest technology, armored with iron plating, hulls reinforced with cross planking, and berths warmed by ducts from coal stoves, bearing all the outward confidence of Britain's aim to tame the most remote parts of the globe.

What Franklin did not know, as his ships snaked through the narrows between Prince of Wales and Somerset islands, was that the caprice of Arctic weather would compound errors of provisioning and set their entire mission on a course for disaster. Ice-locked near King William Island for nearly two years, Franklin's crew would bide their time, refusing to adopt indigenous solutions to combat scurvy and eating from tin cans sealed with lead that leached into their food, possibly poisoning them. They perished under harrowing circumstances, from starvation and cannibalism, as inferred from Inuit eyewitness accounts and piles of human bones on King William Island bearing the telltale signs of butchery and cooking. The

disappearance of the Franklin expedition, and the abject failure
of over thirty-five rescue missions meant to find them, cast a
long shadow that soured the imperial interests of the British in
the Arctic.

But before all that, in a late-summer moment, Franklin most
likely would have seen bowhead whales. The species had already been decimated by over two centuries of wanton whaling
off the coast of Greenland, but they were still present in the far
reaches of the Arctic, where whalers would not ply. I imagine
Franklin burying his chin in the high collar of his naval great-
coat, pausing on the quarterdeck to watch two bowhead whales,
perhaps a cow and its calf, surfacing out of the aquamarine
mirror. The blows of the mother would have towered to the
level of the ship's deck, clouds of water droplets hanging momentarily in the cold air. Perhaps Franklin and his officers
caught sight of the whales' dull black bodies and ivory chins as
mother and calf rolled near the side of the *Erebus*. After a few
moments, the short encounter would be over, and the two pairs,
cetaceans and ships of men, would resume their individual fates
in different directions.

Fast-forward to the start of the twenty-first century, a thousand
miles westward, off the northern coast of Alaska. On an over-
cast spring day, several dozen native Alaskans—Iñupiat women
and men, young and old—pull a heavy, braided line across a
shoreline of land-fast ice. The effort requires every able-bodied
person in the town. The taut end of the rope is wrapped around
the tailstock of an immense bowhead whale, a cow harpooned
by whaling captains in small boats only hours earlier in the
Chukchi Sea. Just like their ancestors, the group succeeds in

the unbelievable task of hauling a fifty-five-foot-long mammal carcass out of the water onto solid ice. At the other end, blood drips from the great arched mouth of the whale. The long baleen plates in the mouth are askew, like improperly folded window blinds. After a moment of silence, followed by a few words, one of the whaling captains climbs to the top of the carcass. Without much fanfare, he directs his crew members to begin the careful, arduous process of removing blubber from flanks of the bowhead's body, peeling back foot-thick sections of skin and fat from muscle using large hooks and flensing knives.

The blubber, tail fluke, pectoral flippers, and some long back muscles are all set aside, in piles designated for whaling crews, families, and everyone in the community. Parts of the carcass will feed an entire household for months. Red-cheeked children running about with unzipped jackets pause in their games to marvel at the whale and press their fingers into the rubbery black skin near the head, leaving temporary, tiny divots.

But no one here knows the whole story of this particular whale, which was a calf when it encountered Franklin and his ships. At two hundred years old, eras of human history are bookended by her lifetime. She was born before the age of coal, when whaling ships had sails, and escaped the devastation of Soviet whaling from massive diesel-powered ships with explosive harpoons more than a hundred years later; she ingested radionuclides from atomic testing, widespread in marine food webs after World War II; and then she experienced myriad sonic changes to her underwater habitat from oil exploration, cargo shipping, and military sonar in the past fifty years alone. Over her lifetime, she gave birth to several dozen calves before falling

to a twenty-first-century Iñupiat harpoon, as many of her relatives did more than a century ago. Lining up the chronologies
together, it is possible that the lifetimes of three generations of
bowheads—grandmother, mother, and twenty-first-century
calves—span more than half a millennium, from Basque whalers
sailing galleons west of Greenland in the time of Shakespeare
to a twenty-third-century Arctic Ocean no longer synonymous
with ice. Bowhead whales are time machines unlike any other
mammal on the planet.

Though the scenarios described above are imagined, the basic
outlines of this story are fact. Bowhead whales are the only baleen whale species that spends its entire life above the Arctic
Circle. Bowheads are known as the one true polar whale for
good reason—they alone have the size and strength to deal with
the vicissitudes of ice, including the wherewithal to break it up
should it suddenly begin to close up around a breathing hole.
(Many other Arctic whale species, such as narwhals and belu-
gas, perish under these circumstances.)

In the summer season, most bowheads migrate from the
Chukchi and Beaufort seas off Alaska east toward the Canadian Arctic, some to the very same embayments where the *Erebus* and *Terror* now rest on the ocean floor. By late fall, most
then migrate west, along the Alaskan shoreline. Some pass
through the Bering Strait into the Bering Sea, while others hold
position north of the strait until they return east at the turn of
the season. This annual cycle forms a figure eight pinched by
the winter pack ice spreading down from the North Pole, forc-

ing large numbers of bowheads to pass relatively close to the northern coastline of Alaska through openings in the ice, called leads. Consequently, sea ice dictates the timing of Iñupiat hunts, which have been part of their subsistence culture for thousands of years.

In general, whales are like most mammals in that they grow rapidly in the first years after birth until reaching adulthood and then slow down, adding little in the way of length for every added year of growth. With access to abundant carcasses, researchers at whaling stations in the early twentieth century counted pregnancy scars on the ovaries of whales and plotted these numbers against total length as a way to gauge the timing of sexual maturity. (No similar technique exists for males, which lack ovaries.) They found that other large baleen whales, such as blue and fin whales, show remarkable growth profiles, achieving sexual maturity in fewer than five years, a rate implying the addition of over a hundred pounds of growth per day. By contrast, the first studies looking at similar growth profiles for bowheads showed that they matured more slowly, taking nearly two decades, on average, until their first pregnancy, not unlike us.

Aging a whale is not trivial or straightforward. For toothed whales, a single tooth sectioned lengthwise reveals annual layers, like tree rings. For baleen whales, researchers have to use other so-called accretionary tissues, such as baleen and ear plugs, to guess age. Like fingernails or hair, baleen provides only a record of growth tied to the longevity of the tissue, which wears down and is replaced through the course of life. Bowhead baleen records a duration of time on the scale of years to a decade, though certainly not a lifetime. Baleen whale ear plugs—accumulations of wax inside the head that don't connect to an external ear opening—show layering that seems to correlate tightly with a whale's length. The only problem for bowheads is that, for an unknown reason, their ear plugs do not have layers to be read.

In 1992 a fifty-one-foot-long female killed off Barrow (now officially known as Utqiaġvik), Alaska, provided much clearer evidence of bowheads' longevity. Having stripped it of its blubber, scientists saw above its shoulder blade what looked like an old injury. "We followed the line of scar tissue and sliced into a big, obvious pocket until we heard a crunch," Craig George, resident biologist in Utqiaġvik for more than thirty years, told me. They discovered a stone harpoon lodged deep in the bowhead's body. Flint and slate harpoons fell out of favor with native Alaskan whalers by the 1880s, thanks to metal versions introduced by whalers in the 1850s, when Alaska was still Russian territory. Considering that the whale must have been relatively mature at the time to survive such a strike, Craig and his colleagues surmised that the whale, labeled 92B2, was at a minimum about 130 years old. When they later counted the pregnancy scars on her ovaries, the age that they calculated was completely consistent with a healed wound from the nineteenth century: 133.

Craig and his colleagues more commonly use another technique to age bowheads, capitalizing on fresh tissue from Iñupiat hunts: eyeballs. The crystalline lens inside the eyeballs of almost every vertebrate animal is composed of proteins that change in chemical structure over an organism's lifetime at a consistent and known rate, keeping time like a clock, in a process called racemization. Elsewhere in the body, cells replace these proteins, resetting the clock; but those in the lens have been there since birth, bottled off from the body's circulation. Lenses thus make great biological chronometers, but only if you can find them before decay sets in.

When Craig and his colleagues studied a sample of eyeballs collected over several years from recent bowhead hunts, the ages that they calculated confirmed the true longevity of bowheads, with many individuals well over a century in age. The ages that they calculated set records not only for whales but for any known mammal. In fact, the oldest whale in their data set, a forty-eight-foot-long male killed in 1995, was an astonishing 211 years old. In other words, it is entirely possible that bowheads born during the time of the Lewis and Clark expedition still swim in the seas off the North Slope.

These multiple lines of evidence—from harpoon technology, pregnancy scar counts, and protein racemization—lend tremendous strength to inferences about the longevity of bowheads because these facts, as Craig remarked to me, "would have seemed ridiculous if you only otherwise had one clue." (When he told his mother, children's book author Jean Craighead George, that bowheads could live more than two hundred years, she wrote *Ice Whale*, about a whale encountering many generations of a single clan of native Alaskans.) Bicentenarian whales carry many stories in their bodies, telling us about the past worlds of the Arctic. You just have to know where to look.

One great place to look for stories is in a whale's mouth. Baleen is not as perfect a chronometer as eyeball lenses, but it is an accretionary tissue that grows in layers, and thus can reveal something about the life of its owner. Baleen plates wear down while in use over a whale's lifetime, but imagine for a moment you could line up plate after plate, decade by decade, over calendar years—with enough baleen plates, you could create a record over centuries. A lot of scientists working on bowheads have had the same thought; they just need enough baleen plates.

In the great warehouses of the Smithsonian's off-site collections, baleen plates of bowhead whales are housed in a series of unusually wide metal cabinets eight feet long; at their longest, the plates can reach up to fourteen feet, though most in the collection are no more than five feet long. Baleen becomes frayed with use and entangled with tubules from adjacent plates moving forward and backward on one side of the mouth, like Velcro, to form a dense mat. In life, this tangle forms a sieve on the right and left sides of the mouth to trap incoming prey—a filter for a filter feeder.

You can think of baleen as a kind of record-keeping device mounted in a whale's mouth that accumulates a sample of its environment for many years until it wears away. In that way it is a lot like hair or really long fingernails, keeping a record of an organism's growth for as long as the structure holds together. The longer the baleen plate, the more time recorded from that whale's life.

At a molecular level, the composition of baleen acts as a kind of tissue archive, recording subtle differences in the ratio of different weights of carbon, nitrogen, oxygen, and other atoms consumed by the organism. With each layer of growth, these atoms (and the ratios in which they were formed) are deposited in baleen, which grows with the seasons—the cyclical bands of growth, like tree rings, show seasonal fluctuations from bountiful times during summer feeding to fasting during migration. Bowheads have absolutely the longest plates of any baleen whale, much longer than those of even blue whales. Lacking any other kind of time-stamped recording device for this species, the layers preserved in a baleen plate are the closest that we can get to annual snapshots from the past life of an Arctic time machine.

Pull out one of the drawers from the baleen cabinets, and this history is literally spelled out on these specimens. In some cases it's stenciled on the plate in tall, angular print by Yankee whalers, or written on an edge-worn specimen tag. The oldest specimen in the Smithsonian has one such tag, inked in oversize cursive handwriting, stating simply: "Arctic Ocean, 1840."

Sampling baleen from the early-to-mid-nineteenth century captures environmental signals from a world before the widespread release of carbon dioxide from industrial fossil fuels into the atmosphere. Since that time, burning fossil fuels has added isotopically lighter carbon into Earth and ocean chemical cycles, adding a distorting layer. Interpreting the signals from a decline in carbon isotope ratios in baleen over the past 170 or so years is tricky and depends on understanding both climate cycles and food webs. This decline could be the result of longer-term background environmental change (part of episodic cycles driven by dynamics in Earth's orbit, which set oceanic cycles that last centuries or millennia), or it could signal something biological, related to the loss of productivity at the base of the food web in the Arctic. Preindustrial bowhead baleen can give us a glimpse of a deeper view of the oceans, before the world changed with the widespread burning of fossil fuels.

Bowheads have always had longer life expectancies than we do, but now, within the span of one human lifetime, we are creating changes to the Arctic that mean any bowheads living today do so in a liminal gap between a familiar past and a potentially unrecognizable future. The Arctic is warming twice as fast as the rest of the planet: the Arctic that existed when I was born was more similar to the one that Franklin saw than it will be to the one my children will know. The most obvious indicator of this change has to do with ice and its different incarnations—from seasonal ice rafting on the sea to older ice formed over several years to near-geologic glaciers. Each year, the maximum extent of winter sea ice has steadily dropped, losing more than half its geographic extent and three quarters of its volume since I was born in 1980. Even older sea ice, which was previously resistant to multiple summer melts, has disappeared. This decline has a distinct downward trend toward a threshold, perhaps as soon as the mid-2030s, below which a summer in the Arctic will be mostly free of sea ice.

Then there are the glaciers. The landmass of Greenland holds the greatest amount of glacial ice on the planet after Antarctica, and it's disappearing rapidly. The amount of water locked in these ice sheets is nearly incomprehensible: adding the volume of water trapped in Greenland's ice to the world's oceans would raise the global sea level by more than twenty feet. Other key Arctic components undergoing major changes, such as the warming of the permafrost, or tundra soil, independently point to the fundamental unraveling of a unique Earth system, in a way that has not happened since the first glaciation of the Arctic more than three million years ago, a time before the first bowheads. A bowhead calf born today will live in an Arctic that, by the next century, will be a different world than that experienced by all of its ancestors.

A warmer Arctic represents a complex fortune for ice-loving whales. First, an ice-free summer in the Arctic means a less impeded Arctic Ocean, open to increased shipping. Rather than being choked with ice, the Northwest Passage will become an express lane for container ships and oil tankers from Asia to North America and Europe—a faster route than through the Panama Canal. More traffic in the Arctic means the unavoidable realities of oil spills and ships striking whales. If the *Exxon Valdez* disaster is any guide, an oil spill in remote parts of the Arctic would be a calamity even harder to contain. Ship collisions are a major threat to right whales, the massive nearest cousins of bowheads, whose numbers never rebounded from whaling in centuries past. There are three species of right whales, and they collectively total only a few thousand individuals across the world's oceans. Their vulnerability is partly due to yearlong gestation periods producing a single calf, with about three to four years between pregnancies. Until 2004 no one had documented a North Pacific right whale calf off Alaska for a hundred years (and not for lack of effort either). Right whales' summertime feeding patterns already take some of them directly into the harbors of major cities, such as Boston. Like bowheads, right whales are simply too slow to avoid collisions with incoming ships that can be hundreds of feet long and move ten times faster than the whales can swim.

A warmer Arctic also creates ecological opportunities for invading species. Killer whales, for example, typically avoid the ice-locked waters that bowheads can break up with the high arch of their snout, but a less icy Arctic has made killer whale visits to this part of the globe more frequent and persistent. Rake marks from killer whale bites are not uncommon on bowhead fins and flukes; other Arctic whales, such as narwhals, avoid killer whales as much as possible (though it's unclear whether killer whales kill and eat bowhead calves as they do with other baleen whales).

In evolution, these kinds of sweepstakes events that involve invasions into newly available habitat can have outcomes tilted strongly in favor of one set of species, to the detriment of others. For example, when the Panamanian land bridge finally emerged to link the Americas, dogs, cats, bears, and many other placental mammals from the north were much more successful in invading South America than austral animals, such as possums, moving north. The analogous outcome for whales invading cooler latitudes from expanding belts of warm latitudes is very much unclear. Bowheads have the advantage of incumbency, but their home field is changing within the scale of our lifetime, to say nothing of their much longer one.

Last, bowheads feed on zooplankton just one to two steps

away from the base of Arctic food webs. The loss of ice opens entirely new habitats for bowhead prey by giving primary producers—phytoplankton, unobscured by ice—an extended growing season, which in turn increases the abundance of high-calorie krill and slower-moving, lower-quality copepods, both types of crustaceans that are favored by bowheads. Less ice means more mixing of nutrients in the water column, from the surface to the seafloor, which enhances the overall productivity at the base of the food web. The cumulative effect of these physical changes on the biology of the ocean food webs will transform the Arctic Ocean into a pelagic one—an ocean structured more like what we see at temperate latitudes. For the near term, as far as the food factor is concerned at least, bowheads should benefit from a new ocean.

But it's not that simple. Rising carbon dioxide levels in the atmosphere are driving increases in global temperatures, and the effects of this change will be felt by every organism on Earth, whether in positive or negative ways. The oceans are a dead-end dumping ground for carbon dioxide, which in its dissolved form makes the oceans more acidic. This process has a notable negative effect on any organism that uses calcium carbonate to grow a shell—including krill and other zooplankton that form the primary prey for many species of whales. It may be possible for krill and other shell-forming species to adapt to a newly acidic ocean, but we don't know how fast this kind of evolution might happen, especially relative to the rate of acidification, which may increase in a nonlinear way.

Often scientists talk about climate change with predictions based on trends that progress in linear ways. The challenge

with complex and dynamic systems—those involving cells, organisms, and even whole ocean current systems—is that they
can respond in large leaps, or shifts, from one trend to another,
suddenly and with little warning. These tipping points are
moments when systems previously buffered against dramatic
change finally flip to a different trajectory. Oceans and climate
systems have changed abruptly in the geologic past—the question for those living in the Anthropocene is not whether these
tipping points exist but the factors that initiate them, and how
soon they might happen.

Bowheads preceded humans in the Arctic: skulls and jaws from
ten-thousand-year-old bowheads have been found throughout
the islands of the Canadian Arctic Archipelago, eclipsing the
oldest evidence of indigenous whaling cultures. With weathered
elongate arches and swooping curves, ancient bowhead skulls
and jawbones that litter the barren shorelines could be mistaken
for shipwrecked timber. As for Franklin's two ships, they now
rest on shallow seafloor near King William Island, but his expedition left a scatter of clues, including frozen bones and a single
note cached in a stone cairn. These are relics of an era now past,
and not just the era of imperialism or exploration. Immense
systemwide changes to the icebound Arctic are unavoidable, at
least through the mid-twenty-first century, barring large-scale
geoengineering that would reverse the loss of ice.

Whales have interacted with humans for tens of thousands
of years, but only in the past few hundred years have such en-
counters affected their destiny. The first changes of any magnitude

happened with the devastation wreaked by industrialized whaling. In the past half century, pollution—both material and acoustic—along with other side effects of industrialized civilization, such as ship strikes or fishing gear entanglement, present clear threats to bowheads and every other whale. There are centenarian bowheads alive today that have witnessed all of these changes; as the Arctic unravels within the scale of our own lifetime, some of these bowheads may still outlive us, reaching a bicentenarian age in an Arctic Ocean with far less ice and many more humans. Today the all-encompassing influence of climate change represents the greatest human impact—in magnitude and rate—that we have yet had on the lives of whales, or any other creature on Earth. Who wins and who loses in this new world of our making remains to be seen.

14.

SHIFTING BASELINES

What an organism eats means everything for its place in the economy of nature. Pull any ecology textbook from before the twenty-first century off the shelf, and you'll see enshrined the graphic of a trophic pyramid. This pyramid, borrowed from the same visual lexicon that gave American consumers the food pyramid, shows a wide base, followed by smaller, successive steps up the triangle, with a point at the top. These stages are meant to represent the way that energy flows through ecosystems in the natural world, and the corresponding way that biological investment—the collective mass of all the organisms using that energy—winnows at each step. This graphic is a convenient and intuitive way to talk about two important ideas underpinning ecology. First, it shows that energy flows through food webs; second, it compartmentalizes organisms' roles in a hierarchy, showing the relative number of players at each stage.

In a trophic pyramid, the base layer represents the primary energy fixers of the world, including organisms that can harvest sunlight using photosynthesis, such as plants or, far more abundant in the oceans, phytoplankton. The next layer, above the primary producers, is composed of primary consumers, such as

zooplankton, which feed directly on the sun harvesters, followed by successive stages of consumers until crowned by the
top consumers—mainly big, charismatic vertebrates but also,
arguably, humans. We are the biosphere's top consumer.

Biomass in marine ecosystem pyramids are actually built
differently from those on land. While trophic pyramids on land
generally have the stepped structure of a narrow, stacked pyramid with a very large base, marine pyramids are slightly inverted at the base, with far more biomass in zooplankton than
with phytoplankton. This difference is largely about the high
rate of turnover in phytoplankton—at any one instant, organisms at this lowest level are not as persistent in the environment
(that is, long-lived) as those one level up, and thus don't count
as much for instantaneous biomass in an ecosystem.

As zooplankton, krill sit squarely on the second level of a
trophic pyramid—one step above phytoplankton, which con-
vert sunlight into biomass. This ecological organization is the
reason that scientists describe baleen whales as feeding merely
two steps away from sunlight—as long as they have the means
to feed efficiently on this part of the trophic pyramid (using
baleen), they can capitalize on the greater abundance of prey
(by numbers) and minimize energy lost at higher levels. Large
baleen whales skip up the pyramid as top predators because
there are essentially no creatures that kill them, besides the rare
times a pod of killer whales decides to go after an adult. Krill-
eating whales thus might not be apex predators in the way that
killer whales are—but baleen whales are still fairly character-
ized as major ocean consumers.

For decades, studies pointed to an overwhelming argument

that the boom and bust of primary producers in the ocean (phytoplankton) have a direct correspondence to the occurrence of whales, in timing and geographic space. In other words, whales follow their food in the ocean. Plankton aren't distributed evenly across the oceans; instead, their presence is dictated by large-scale oceanographic processes, such as upwelling. Ecologists therefore argued that everything about the ecology of whales was controlled from the bottom up—that lower trophic levels had a deterministic effect on top levels.

But there are also top-down trophic interactions that cut against the exclusivity of the bottom-up view. A classic example of top-down trophic mechanisms involves kelp forests off the Pacific coast of the United States. Sea urchins devour kelp fronds, and sea otters are very fond of eating sea urchins. Scientists recognized the impact that sea otters made on the physical structure and expanse of a kelp forest only when sea otters were reintroduced to the Pacific coast after over a century of overhunting, mainly for fur. In places where sea otters were reintroduced, kelp forests rebounded, having been released from the ecological constraint imposed on them by sea urchins.

Now add killer whales into the picture. In 1998 the marine ecologist Jim Estes and his colleagues argued that prey switching by killer whales in southeastern Alaska had unforeseen but clear ecological effects on lower trophic consumers and producers, such as starfish and kelp. In the field, Jim and his colleagues observed killer whales eating sea otters instead of seals and other marine mammals—leading him to wonder about the implications of a dietary equivalent of making lunch out of popcorn instead of an all-you-can-eat buffet. Their argument was pulled from the same logic as the original study on sea urchins and kelp forests: a dietary switch by an organism at the top of the trophic pyramid had effects that echoed down to the bottom levels. Their proposal led to widespread debate among ecologists about the prevalence of top-down versus bottom-up interactions in food webs. Much of the debate, however, didn't consider recent history—in this case, an outstanding question of what killer whales ate *before* whaling in Alaska (and other parts of the world) fundamentally reorganized the number of big consumers in the oceans, including many large whale species that likely served as prey.

This question about what happened in marine food webs—what killer whales did before whaling, before the discipline of ecology even existed—underscores perhaps the most important idea in ecology today: we can't assume that the size of animal populations we see today have always been this way. Shifting baselines is an idea to describe our collective cultural amnesia about how the world once was. This kind of amnesia happens when we try to measure a system undergoing massive degradation and in the meantime forget the location of past goalposts, leading to consistently shifting measures of normalcy, from generation to generation. Fishery scientists first applied the term of shifting baselines for the real phenomenon of diminished expectations of fish size or yields from overfishing. The result over the years was a dramatic shift in the concept of what would be normal for a fishery—smaller and smaller fish, and fewer and fewer of them. The idea has since gained broader traction among conservation biologists because of its utility in describing any ecological system affected by humans. It applies as

much to passenger pigeons and bison as it does to whales, because no one alive today remembers the full scope of what the
baseline abundance of these animal populations was once like.

There are ways, however, that we can infer what these very
recent whale worlds looked like. When there are enough DNA
samples available—which is the case for some species of whales,
such as humpbacks—scientists can begin to use sophisticated
methods to infer what their standing genetic diversity means for
the history of their lineage. Humpback whales are among the
many baleen whale species where we expect to see signs of genetic bottlenecking because of whaling. At low population
sizes, the detrimental effects of insufficient genetic diversity can
take hold (inbreeding as one example) and leave a genetic signal
that lasts for generations. With some assumptions about muta-
tion rate and knowledge of existing population size, scientists
can then estimate population size at different points in the history of a group. One of the startling outcomes of this work in
humpbacks was the inference that humpback whales were many
times more abundant before whaling than they are today—
some six times more, according to the results—a figure that
conflicts with the only other source of historical data available,
whaling logbooks. These latter records provide tabulations of
historic kills, but the results from the genetic diversity studies
seem to say that logbooks are not telling us the whole story
by underestimating prewhaling populations by several magni-
tudes. It is hard to discern a likely historical value between these
two sources of data, but, if true in any magnitude, the comparison tells us that the depleted world of whales today may be
missing much of the ecosystem function and productivity that

supported so many more whales only decades or centuries ago. The idea that whale baselines have shifted in the time that we've studied their ecology gives us a new way of looking at what we thought we knew about many species and their lives.

Much of the knowledge that we accept as foundational about ecosystem function is born out of fieldwork on a biosphere that has been severely altered by human activity and the resultant removal of huge amounts of biomass. The challenge for any whale ecologist is to understand baselines—and even whether such data matter or exist in the first place—relative to the question at hand. For example, did whales strand in different numbers (or different ways) when there were many more of them? Or what about those whalefall communities in the deep sea—what were they like before (and after) whaling removed hundreds of thousands of potential carcasses that would have otherwise rained down to the seafloor?

The ecological questions about energy use up and down food webs tend to mostly involve questions about organisms eating one another. But organic waste also factors into the equation. Yes: whale poop matters to the ocean ecosystem, on a significant scale. Whale feces aren't particularly solid; they tend to be fleecy and float at the water's surface until they fall apart. As they disaggregate into the water column, they bring nutrients to the surface that were previously sequestered much deeper, until a whale ate them at depth and then expelled their remains much farther up in the sunlit zone. Rarely, sperm whales will expel masses of undigested squid beaks that float and hold together like rotten mulch rolled into a ball. These particular feces are called ambergris and were once prized by the perfume industry for the pungent, sweet smell they exude—smoky and almost familiar, like a relative you last saw as a child.

In the oceans, zooplankton like fish keep a cycle of nutrients (mainly compounds with nitrogen) suspended in the photic zone, until their remains—fish bones, plankton exoskeletons, shells— rain down on the seafloor in the form of tiny particles of biological debris called marine snow. (Eventually, millions of years later, these remains may return to the surface via tectonic uplift.) Generally scientists call the processes that keep nutrients moving through the ocean a biological pump, because they move biological productivity across different depths in the water column. But add whales to the picture—especially at the scales of their prewhaling abundance—and suddenly the role of these large consumers becomes important for biological pumps in the ocean. Unlike fish and other zooplankton that migrate up and down in a narrow band of the photic zone (in step with daylight), whales transport tons of nutrients from much deeper depths all the way to the surface, where they need to breath and, as it turns out, poop. The act of defecating at the surface, usually after feeding, essentially fertilizes the entire food web, enhancing further productivity and nutrient cycling by other zooplankton. When biologists calculate the amount of biomass that prewhaling abundances of whales add to this system with their feces, the total amount of nitrogen input into a local food web can exceed that nominal amount brought in by nearby rivers and the atmosphere. Whale poop can supercharge a whole ecosystem.

Consider, for a moment, what happened to these processes during the course of industrial whaling. The systematic removal of living whales—their flesh for meat and oil and their bones, ground to bonemeal or simply dropped overboard—would hinder productivity of these nutrient cycles. We know that we are living in a less abundant world after whaling—are we somehow living in a world with a depressed ecological function too? That remains an open question. The more we discover about the important roles that whales play in ocean ecosystems, the more it's apparent that the legacy of whaling has far broader consequences than we might have originally imagined.

15.

ALL THE WAYS TO GO EXTINCT

I awoke in a strange bed, momentarily disoriented. When I heard howler monkeys calling in the distance and opened my eyes to see netting over my bed, I remembered: Panama. We had left Panama City for the Caribbean coast late the previous afternoon. I got up and gathered with my colleagues for a breakfast of eggs, papayas, and coffee set out by our hosts at the small, spare hostel before we set off in pickup trucks.

Jorge Velez-Juarbe and I had come to Panama to collect a fossil whale skull that a local university student had noticed but wisely photographed and left in place. When the student's adviser, a Smithsonian scientist, asked me to come down to Panama to help excavate it, I didn't hesitate. I knew Jorge, then a graduate student and today a museum curator, would also be game; having grown up in Puerto Rico, he's also an expert in the paleontology of the Caribbean. Based on photos, we could see a fragmented snout bearing heavy, pointed teeth poking out of rock. Surrounding it, an outline of the skull's braincase, which looked to be safely hidden in sediment. From the teeth alone, I hazarded that it might have belonged to a common fossil species found in Europe and North America, although I couldn't be sure. It didn't matter, though: the fossil record of whales from the tropics is meager, and anything as substantive as a skull would be important. The only catch: the tides on this stretch of Caribbean coast limited access to the ledge where the skull rested to only one day in the calendar year. We would have a matter of mere hours to exhume and safely wrap the fossil.

Getting to Panama was relatively easy, at least. The Smithsonian has a deep connection to Panama, braided with the geopolitical history of the country and originating with the canal. The man-made waterway that cuts across fifty miles of land, connecting the Atlantic and Pacific oceans, remains one of the great engineering feats in human history. The United States had negotiated near-sovereign rights to build the canal as a strategic outcome of assisting Panama's separation from Colombia. Soon after construction began on the Panama Canal in 1904, Smithsonian scientists dispatched to control insect-borne diseases in the Panama Canal Zone documented undiscovered flora and fauna of the isthmus. Realizing the scope of biological richness to be studied, the Smithsonian successfully maintained a footprint in the country, which has since expanded to half a dozen field stations today, reaching from ocean coasts to tropical forests and comprising the Smithsonian Tropical Research Institute. STRI's field stations are more than just shelters for working scientists: they are hubs of scientific inquiry, serving as platforms for the careers of thousands of scientists from all over the world.

For scientists interested in Deep Time, however, Panama is synonymous with its isthmus. The rise of the S-shaped spine that makes up the country is a pivotal moment in recent Earth history. Oceans swept east to west for over 150 million years, without any kind of land bridge between North and South America. The narrow bridge of land that makes up Panama today is the result of a set of tectonic collisions generated from a line of volcanic activity and mountain-building episodes, which culminated in enough land being pushed upward and eventually out of the water to fully sever any equatorial communication between the Pacific and Atlantic around three million years ago.

As our truck rambled through a hilly landscape of tangled green forests and farms, road cuts along the way revealed sandy, gray marine rocks I knew to be older than the land bridge, uplifted by the continued tectonic thrusting. Jorge snapped photos and reflected on how the isthmus changed the world—altering both land and sea ecosystems at the same time. Terrestrial species, even lumbering ones like armadillos or giant ground sloths from the south and bears and camels from the north, were able to wander at will. For marine species, the land bridge meant a complete reorganization of global ocean currents. It severed gene flow and separated evolutionary paths for bryozoans, clams, reef fishes, and probably even whales—although some species of whales would have migrated around Cape Horn or even the Arctic, though only during ice-free interglacial periods.

We arrived at the small town of Piña just after noon and followed GPS waypoints toward the fossil site, ambling down narrow ledges overgrown with vines and brush. Tides easily change the complexion of exposed rock, so it took some time to find the precise location of the skull. Under the high equatorial sun, Jorge and I worked at the wet, sandy rock, nearly nose to nose, being careful not to strike the skull—or each other. Usually we would have taken more time, mapped out the bones along a grid, but we were racing against the tides. Years of anatomical training gave me a guide for the location of bones that lay beneath, but you honestly never know until you start digging—carefully, but in this case as quickly as possible. As we dug deeper, we discovered a set of jaws beneath the skull; my hammer stroke had left a clean break in the bone, much to my chagrin. "That's why glue exists," Jorge joked, a common refrain. We applied some acrylic glue mixed in the field and continued our work.

Watching the incoming tide, we managed to tunnel under the entire block of rock containing the skull and jaws, wrapping the coffee-table-size block in spool after spool of plaster bandages. We hauled the block and our equipment higher up on the shore and took a few moments to rest, the first pause after four hours of furious work. The plaster had barely cured by the time the excavation pit was underwater. Tomorrow's tide, like every other one for the rest of the year, would be at the wrong time or height to permit a return.

Exhausted and sweaty, we held up chalky hands to each other as we smiled and labeled the plaster block with all of the appropriate landmarks to orient it back in the lab ("This side up"), a tracing of the general outline of the skull and jaws' position, and a note of where to open the plaster cocoon ("Cut here"). We wouldn't know until months later, but we were wrapping up a species new to science, which ended up telling us as much about extinction and river invasions in cetaceans as it did about the isthmus of Panama.

Many species of whales, even large baleen ones, occasionally migrate up big freshwater rivers, but the vast majority of them are temporary visitors, inquisitive or just plain lost. Only a few lineages of true river dolphins exist, with the required specializations needed to live in freshwater rivers that run hundreds and even thousands of miles away from the ocean. Getting to their habitats requires muddy-boots science—forays inland through forests, streams, and countrysides. Their obscurity and the increasing rarity of these lineages means that few people actually get an opportunity to see wild, living river dolphins in the Ganges or Amazon, some of the last places where they still survive.

We have a good sense of the range of adaptations that whales need for freshwater life because the transition from seagoing to riverine habitats has happened several times, across several different continents. Each lineage could be thought of as an evolutionary experiment; the traits they share tell us about the broader evolutionary solutions required for whales to adapt to freshwater habitats. For example, unlike almost every oceanic whale, river dolphins have low dorsal fins and highly flexible necks. They have broad, fanlike flippers instead of scythe-shaped ones. They also have strikingly small eyes; one species, the South Asian river dolphin, has hardly any eyes at all. Scientists suspect this latter trait has to do with the murkiness, or high turbidity, of rivers with suspended sediment; it means less of an emphasis on vision, which is fine for an echolocating mammal. Echolocation is akin to seeing with sound—imagine toothed whales possessing acoustic flashlights whose beams get focused with a turn of the housing. All of those beams scan the external environment, returning echoes of whatever is out there, whether inanimate, desirable (like prey), or to be avoided (like predators).

What is perhaps surprising is that the solutions for living in freshwater don't seem to require drastic changes. Whales haven't had to evolve anything new; they've just had to emphasize or refashion existing traits. Another peculiarity is the fact that river dolphins have returned to the same freshwater systems that their ancestors inhabited some fifty million years ago, during the time of *Pakicetus*.

Parsing the question of how whales ended up adapted for life in freshwater leads back, in the end, to a sole specimen. To gain an understanding of any species or lineage, scientists must first start with such an individual. These single specimens are called types, and they are the ne plus ultra primary reference material— the specimens to which any other possible specimen from that species must be compared. More practically, type specimens need to be held in the custody of a museum, where any scientist can go see them, study them, measure them, and—in some cases—even destructively sample them.

Types matter for species that have gone extinct because they are the touchstones for *any* kind of biological question—once the last individual has died, there's no way to increase the sample size, measure changes that happen with age, or study the variation in a trait over generations (the fuel of evolution). The last known individual Yangtze river dolphin, *Lipotes vexillifer*, died in 2002 in captivity. No one has seen or heard a Yangtze river dolphin since then; the last survey in 2006 failed to find any.

The type specimen of the Yangtze river dolphin rests in a case in the west wing of the research collections at my museum, but it was collected over a hundred years ago by Charles Mc-Cauley Hoy, a teenage hunter hired by the Smithsonian to collect bird and mammal specimens in the Hunan Valley of China. Like many natural history museums eager to expand their collections from remote parts of the world, the Smithsonian hired locals with a particular blend of wherewithal and ambition: as the youngest son of missionaries, Hoy knew the hinterlands of the Chinese countryside. He had recognized that there were two different kinds of cetacean species in the lake that flowed off the Yangtze River: the baiji, the common Mandarin name for *Lipotes*; and a smaller, finless porpoise, which Hoy called "the black species," now known by the scientific name of *Neophocaena asiaeorientalis*. (It teeters on the brink for reasons that similarly drove the baiji to extinction.)

If we want to know what the baiji was like before it disappeared—how it moved, what it ate, where it went to breed—before the handful of surviving members of the species were transferred to concrete aquariums, we can read Hoy's account, beginning on a winter's day in 1916, when he and his boat companions on a duck hunt came across a school of baiji, a species he had long desired to procure. The only published black-and-white photograph of the day shows Hoy wearing a flat expression under the brim of his broad hat—suppressed pride or perhaps a studied pose—kneeling with rifle in hand behind the three-hundred-pound carcass of the future type specimen, its mouth propped open by a stick. His account tells us they swam in pods of ten to fifteen individuals, leaving the water muddy from stirring up the lake bottom to feed on catfish, which Hoy found in the type specimen's stomach. Hoy's report tells us that when the lake's water level rose in the summer, the baiji disappeared, likely heading upstream in other rivers that flowed into the lake, to reproduce. Like the hundreds of thousands of field notes sitting in silent museum archives, Hoy's account of his collecting experience tells us facts that we cannot know today. The world has changed since a young hunter happened to shoot a dolphin in an oxbow lake seven hundred miles away from an ocean.

The crate containing the plaster-wrapped fossils from Panama arrived at the museum about a year later, following the expected bureaucratic delays that frustrate me (and many other scientists) with regularity. At the museum, Jorge and I rolled the crate into the vertebrate paleontology preparation lab, where staff skilled in anatomy, manual dexterity, and material properties started the process of exhuming the fossil encased in plaster and rock. Over the course of several months—fast work, actually, helped in part by the soft rock—they sawed and ripped the hardened plaster bandages away, and then slowly brushed, scraped, and drilled away sediment, layer by layer, around the skull. Sometimes their work involves heavy machinery, like a pneumatic saw; other times, a simple dental pick to scrape away at grains of sand. Here studio art meets laboratory of extinct anatomy. Glue, of course, is always on hand.

Slowly, more and more of the fossil came to light. Jorge and I could see, week by week, the slow reveal of intact bone with skull traits that pointed to a fossil species unlike anything previously reported. I know the disappointment that sets in when a promising fossil turns out to be run-of-the-mill; this was exactly the opposite. Anticipation built as we saw more and more of the delicate and incomplete fossil revealed before our eyes, and we became increasingly sure that it belonged to a species of toothed whale that had been extinct for millions of years. For museum scientists, skulls are far more than trophies—although they do occasionally have that type of glamour. Skulls house organs that control the most important functions in the life of a whale: how it feeds, sees, hears, and thinks. These organs tend to reflect adaptations tailored to a whale's ecology (living in turbid waters, for example). Consequently, skulls tend to bear many of the most distinctive features identifying a species.

The Panama skull was fragmentary, too fragile and too ungainly to hold. Using the techniques from Cerro Ballena, Jorge and I scanned the Panama skull, which provided the basis for a digital 3-D model and then a full-size 3-D printout in plastic, which we could examine and manipulate without fear or limitation—although we kept the original on hand, in a special cradle, for close-up details.

The Panama skull's exposure to the tides had eroded much of the braincase and its left side. We were at least in possession of the top of the skull; it looked reminiscent of the baiji, when we placed them side by side. (Hoy brought the skull and a few other bones from his baiji carcass to the Smithsonian in 1918.) Hoy's baiji skull had an irregular golden color, as if it had been carved from wood; both it and the Panama skull had long snouts and relatively conical, pointed teeth. We were also struck by the similarities of the pedestal-like prominence at the highest point, behind the bony opening of the nostrils.

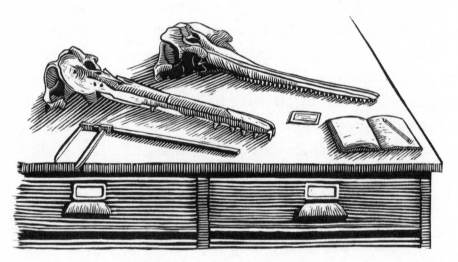

Isthminia panamensis and Lipotes vexillifer

On closer inspection, the configuration of bones didn't quite match, and we decided to expand our comparisons to the skulls of other river dolphins. The problem that we contemplated had in fact vexed students of river dolphin osteology for many decades: are the similarities among riverine dolphin species indicative of shared ancestry or merely convergence in lifestyle? In other words, did they have long snouts because their shared ancestors happened to have long snouts, or did they evolve similar features over time because of selection?

The fact that freshwater river dolphins live in rivers on separate continents would seem like an important clue to their evolutionary history. In the midtwentieth century, most experts downplayed this fact and maintained that the four different then-living species—the baiji from China, one species from India and adjacent countries, and two species from South America—all belonged in one taxonomic group. This classification implied a shared or common evolutionary origin, effectively asserting that all river dolphins were more closely related to one another than they were to any other species, living or extinct. Generally, the idea held that river dolphins were the cetacean version of a living fossil: so-called archaic species that happened to have survived extinction by persisting in freshwater refuges, having descended from a single, widespread marine ancestor. Long snouts were the result of shared ancestry, not similar ecology, in this view.

The evidence from the DNA of the four river dolphin species showed otherwise: they had separate ancestries, and—with the exception of the South American species, which were each other's cousins—they all belonged to different branches on the cetacean family tree. In other words, the world's river dolphins do not form any kind of natural grouping and are mostly unrelated to one another. This molecular finding implied that freshwater lineages colonized each river system on different continents from different ancestors and, possibly, from different times. Long snouts were the result of selection, for snapping up prey in turbid rivers.

With this knowledge, we considered our Panamanian specimen:

we had collected it from marine rocks along the Caribbean side of the country that were between 6.1 million and 5.8 million years old well before the rise of the S-shaped isthmus. Its comparatively large eyes, among other more-nuanced traits, pointed to its having been a marine inhabitant, like modern oceanic dolphins. This similarity was illusory; Jorge and I discovered, through analysis of its evolutionary relationships with other whales, fossil and modern, that its closest relative is today's Amazon river dolphin, *Inia geoffrensis*. The new species, which we christened *Isthminia panamensis*, had little to do with *Lipotes*, of which it was only a distant relative. Instead, *Isthminia* reminded us that features we assumed had evolved for life in the Amazon were also found millions of years earlier in a marine cousin that inhabited the Central American Seaway.

From the farthest reaches of the Amazonian watershed, high in the Bolivian and Peruvian Andes, or thousands of miles upstream in the Yangtze River, river dolphins have evolved a specialized way of living that puts them at high risk for extinction, should anything about those environments change in an unexpected way. Unlike marine species with broad ranges, which might escape such changes by simply swimming elsewhere, the physical isolation of riverine dolphins means that they have limited options when their local environment suddenly becomes hazardous. They have nowhere else to go.

A variety of human activities increases the danger of that dead-end scenario. First, there's wholesale habitat modification: the construction of the Three Gorges Dam—the largest in human history—changed every aspect of the ecosystem in which the baiji had evolved, including its geography, water flow, seasonality, and prey. This scale of all-encompassing change shows how vulnerable any river dolphin is to extinction, especially when you consider the direct ways that humans control waterways, which have inevitably (so far) resulted in more pollution, noise, and bycatch. The corrosive effects of pollution and noise on river dolphin health can be a matter of thresholds and localized concentrations, whereas bycatch is immediately detrimental—the literal though unintended removal of a predator from the system.

The words we use to describe these dangers sometimes trivialize the magnitude of their impact. "Bycatch" sounds abstract, almost like an economic variable, but it accounts for the mortality of over 300,000 cetaceans each year—large and small, common and endangered. One of these species endangered by bycatch is a species of porpoise called the vaquita, known to science only since 1958. They are uniquely found in the northern corner of the Gulf of California, Mexico, and nowhere else. They are among the smallest cetacean species, furtive, and extremely difficult to study. You could cradle one across your open arms, although you would be hard-pressed to: as I write, there are fewer than thirty vaquita left—maybe far fewer, if you talk to the scientists who have looked for them.

For the past few decades, the vaquita's fate has been tied to the totoaba, a large fish that also uniquely lives in the gulf. When stretched and dried, a single totoaba swim bladder fetches several thousand dollars on Asian markets, where they are coveted as gifts, dowries, or, more questionably, medicine. An illegal global supply chain, likely linked to criminal enterprise, is clearly responsible for the unabated demand, which has only increased as the totoaba becomes rare itself. In the gulf, illegal gill nets rake in both vaquita and totoaba indiscriminately; the vaquita is a casualty of market forces both local and a world away.

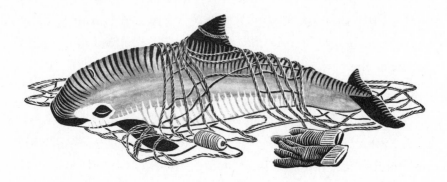

The vaquita

After several years of ineffective plans to limit the scope of illegal fishing (which most directly threatened the vaquita), the strategy to save the vaquita entered a new phase without precedent: capture any surviving individuals and transport them to a sea pen, where they could be protected until the threats in their environment could be controlled and mitigated. The scale of the effort was massive, involving over a hundred scientists from around the world, including veterinarians, behavioral ecologists, acoustic biologists, and fishery scientists. They used an array of listening devices in the water to detect the notoriously shy porpoises and, in an ambitious—and desperate—turn, brought in U.S. Navy–trained dolphins, like sheepdogs, to corral the vaquita.

Nothing like it had ever been attempted before, because the risks with this type of *ex situ* conservation were raised by the vaquita itself: when hauled out of the water, even for a brief period of time, it panics, which means the stresses of capture and captivity can lead outright to death. Unfortunately, that's exactly what happened: after the successful capture and release of a calf, a second vaquita captured by the team succumbed to the stress of its capture. After deliberation, this outcome led to the suspension of the effort, and the grim acknowledgment of the great dilemma with the vaquita: doing nothing will almost certainly lead to their demise; and capturing any living ones and trying to implement a captive breeding program may only accelerate the same fate.

Extinction does have a fixed line in time—set when the last individual of a species dies—but practically, for scientists wanting to measure survivorship for a group of organisms, it becomes more of a probabilistic description: we call a species extinct when we haven't seen it in a very long time. In other words, knowing about extinction is a game of probabilities, not absolutes. Consider any species that has gone extinct in recorded human history. More often than not, it's a story of decline leading to fewer and fewer sightings or captures, until it's just silence for many, many years. In the case of whales, their inaccessibility and huge geographic expanse compounds the challenge in making such a firm declaration about their fate. Even for geographically limited whales—river dolphins or the vaquita—it's fiendishly difficult to arrive at a precise number for a population because it's still not possible to survey every corner and cove of an animal's habitat with either visual or acoustic tools. As a consequence, the status of even the baiji, not seen since 2002, is sometimes qualified as "functionally" or "effectively" extinct as a way to grapple with the possibility (or hope) that we might be wrong. We may never know the last survivor because extinction, especially for animals in the water, happens silently.

Most whale species that ever evolved are extinct—the reality of a fifty-million-year history on Earth. The ultimate reason for their disappearances is imperfectly known. Paleontologists will talk a good game about changing habitats, hyperspecializations, competition, or predation, but the fact is that, lacking a convenient marker such as a mass-extinction event, it is hard to point to a transparent explanation.

Today there is no confusion: human agency, in all of its forms, produces the strongest extinction pressure on whales. The fates of the baiji and the vaquita are likely sealed for different reasons, mostly related to their particular histories and human causes—wholesale habitat modification for the baiji and fisheries bycatch for the vaquita.

While direct hunting from whaling jeopardizes far fewer whales today than it did in the recent past, the indirect pathways of ocean noise, habitat modification, bycatch, and pollution, if unmitigated, represent the primary ways that whale species might go extinct in the near future. There are about half a dozen baiji specimens in the world's museums (including the

Smithsonian's type specimen), and twenty-nine specimens of vaquita accessioned in North American natural history museums. There will be few, if any, additions to these records, which constitute the total material remains of this species that we will ever have. There are whale species for which we have thousands of specimens; and there are some whale species, especially belonging to the family of beaked whales, for which our total biological knowledge rests with a single skull in a museum drawer. For the species that are still fundamentally enigmatic in the truest sense of the word, scientists simply do not have enough information to argue, one way or another, about their ability to survive in the oceans during the age of humans. We need to know what exists if we want to have any hope of preserving it.

16.

EVOLUTION IN THE ANTHROPOCENE

On a skiff off the coast of Alaska, in the inner passage of the Alexander Archipelago, I peered through binoculars at a pair of humpbacks; Jan Straley, a biology professor, stood by the outboard motor. We had tagged the whales a few hours earlier, and we now trailed their slow path through the waterways. Every time they surfaced to breathe, we recorded their cartographic bearing and distance. We had eased into a nice pace following the whales—five blows apiece at the surface, dive for a few minutes, then surface and repeat. Suddenly, midway through one of the blow sequences, the humpback pair abruptly cut it short, dived, and stayed down.

Moments later, I understood why: I caught sight of a tall dorsal fin that can belong only to an adult male killer whale knifing through the air. From far away, the elongate triangle was devoid of reflection, as if someone had cut a hole out of the sky. The tall fin, maybe five feet high, wobbled in its momentary transit out of water and then slipped back through the mirror. Shorter, more rounded dorsal fins, likely belonging to females and younger whales, followed. They moved quickly, surfacing in succession, heading toward the humpbacks and our boat. The pod of half a dozen or so killer whales did not surface again until they left the sound behind us.

A pod of killer whales glides unharried through the ocean because they have no ocean-bound predators. They are the largest species of true oceanic dolphins, and their cunning, curiosity, and other striking behaviors are nearly synonymous with their common epithet, "orca." As the supreme predator of the oceans, their diet includes an entire species list of other marine predators: great white sharks, baleen whales, other dolphins, porpoises, sea lions, salmon, and squid. Killer whales will even pick off the unfortunate land mammal that happens to be paddling in the wrong place at the wrong time, even deer or moose. While humans have hunted and continue to hunt them, only captive killer whales in aquariums have maimed or killed humans. Still, the way the pod moved through the water both surprised and unsettled me. I was happy for the distance between our open boat and the ominous black forms that raced through the sound.

An adult male killer whale might weigh as much as ten tons and reach thirty feet in length. While an adult humpback outweighs a single killer whale by a factor of four, it's the fact that killer whales hunt together as a group that makes the strategic calculus of fight versus flight more complex. Humpbacks have scalloped flippers one third their body length that can be used as weapons in self-defense—the deterrence of a thirteen-foot-long windmill blade on a spinning forty-ton whale. A pod of killer whales will undoubtedly harass adult humpbacks; even though they would probably much rather hunt a calf, they can cooperatively kill whales much larger than themselves.

Jan caught my surprised face. "Oh, that's fun," she offered casually—the cheerful but measured reaction of someone who has seen a lot of Alaskan natural history. "Those are probably transients," she said, referring to the mammal-eating killer whale ecotype. In the Pacific Northwest, scientists recognize three different ecological varieties of killer whales, each genetically and behaviorally distinct. Aside from the mammal-eating one that ranges from Alaska to California, there's a resident group that specializes on salmon runs. Residents are strongly matrilineal and vocalize frequently, whereas transients have pods composed of fluid associations of unrelated individuals that vocalize only after a stealthy hunt for mammal prey that might otherwise hear them coming. A third ecotype—offshore killer whales—are little known, although some studies suggest that they eat sharks. Back in the boat, within a few minutes, the humpbacks resurfaced, showing no outward signs—as far as we could tell—of what had transpired beneath, continuing along their meandering course through the procession of tree-lined islands.

Sitka is perched on a narrow toehold of land between rocky coast and snowcapped mountains in southeast Alaska. Today fewer than ten thousand people live there, but its location on the panhandle of Alaska has long made it a convenient launching point for expeditions, including my predecessors at the Smithsonian who organized the very first ones. At the time, Alaska was still relatively unknown to Western scientists, having been purchased as an American territory only in 1867. Within a decade, Smithsonian scientists passed through Sitka on their way inland, or out along the Aleutian Island chain. The scientists who traveled throughout Alaska in the nineteenth century rarely pigeonholed themselves into the professions we now call paleontology, mammalogy, ornithology, and malacology— they were just naturalists, with omnivorous interests, careful eyes, and assiduous collecting habits. Their work begat collections that continue to inform us about American wilderness near the Arctic Circle. Spending time in Sitka, for me, meant walking in their footsteps. I was there for similarly cross-disciplinary reasons—tagging humpback whales and hunting Oligocene rocks that had already produced scrappy but tantalizing fossils nearby.

On an off day, Jan and I walked over to the Sitka Sound Science Center, a nonprofit housed in a renovated school. Inside, above one of the touch tanks, hung the skeleton of a young killer whale that had stranded on nearby Kruzof Island. Kruzof is a squat, cone-shaped volcanic island that looms quietly in the vista north of town—an inescapable reminder of geologic-scale forces that could spring to life at any moment. It took a coast guard helicopter to haul the carcass, in segments, off of Kruzof's black-sand beaches. During the later necropsy, a pile of bones, hair, whiskers, and claws was discovered in its stomach, a last meal that likely consisted of a harbor seal. Those stomach contents were sure signs that this animal, about seven years old at the time of death (based on later studies of tooth rings), was a member of the mammal-eating or transient killer whale ecotype.

Whereas killer whales once could readily hunt baleen whales, the large-scale removal of this prey resource in the whaling era meant that they had to compensate by eating other available prey, such as seals and sea otters. In a follow-up to their top-down trophic cascades paper, Jim Estes and his colleagues argued that in fact the loss of large-whale biomass following whaling led to a fundamental shift in the diet of marine mammal–eating killer whales in the North Pacific. Jim and his colleagues argued that killer whales progressively ate their way down trophic levels: after large baleen whales were removed, sea lion and seal populations crashed, and then sea otters. Nothing in biology is perfectly clean—especially in ecology at sea—and myriad human impacts likely complicate this simple explanation. Nonetheless, after calculating the caloric needs of killer whales and the energy represented by their prey switching, and scouring the literature for historical data on killer whales attacking baleen whales, Jim and his colleagues found ample evidence to support the idea that killer whale feeding behavior could cause the kinds of ecological effects witnessed in marine mammal populations in the North Pacific over the last half of the twentieth century.

This proposal elicited a lot of debate among marine ecologists because it implied that killer whales, by prey choice alone, could direct strong top-down trophic pressures. Also, the hypothesis relied on events that were difficult to discern: How often did killer whales eat large baleen whales before whaling? How often do killer whales eat sea otters today, even? Ecological interactions, especially for whales, are a matter of careful detective work and inference.

Regardless of what they are eating, killer whales hunt cooperatively, analogous to the way that wolves hunt as a pack. A single killer whale, with its heft and large peglike teeth, should be an unsettling sight for any smaller marine mammal, cetacean or otherwise. But killer whales multiply this effect by coordinating their hunts in pods. Together multiple individuals can attack a much larger organism, even a blue whale, by biting and using their bodies to keep the whale from surfacing (a much easier tactic when their prey are calves or species the same size as or smaller than they are). When their hunts are successful, killer whales will strip meat using their teeth; without opposable thumbs or hands, they need to work together to pull in opposite directions. They then share food among their pod, like many other mammals. Killer whales also have a reputation for killing seemingly for practice or play—catapulting apprehended seals dozens of feet in the air with their powerful tails—behavior that is poorly understood but nonetheless discomfiting to human onlookers.

Killer whale pods also have strong family ties that make them extremely social animals. Generally they are matrilineal, with matriarchs retaining the leadership role of the group—in resident ecotypes, matriarchs rule the entire pod, while relationships are less structured in other ecotypes. In the wild, killer whales can live into their nineties, although in captivity that number tops out at fifty years. There are likely killer whales that lived through the entire rise and fall of industrial whaling, witnessing untold changes to the entire food web both at sea and along the coasts. The chemical history locked inside the teeth and bones of these individual killer whales might say a lot about the top-down versus bottom-up trophic patterns.

Living at the top of a food web for decades also means continual exposure to and concentration of any persistent poisons in your diet. For the same reasons mercury levels concentrate in seafood like canned tuna, killer whales possess some of the most contaminated tissue of any mammal on the planet, carrying high loads of chemicals such as flame retardants and complex organic molecules, which resist rapid decomposition. Like bowhead whales in the Arctic, killer whale pods are mobile archives of humanity's chemical legacy. For killer whale pods that live close to major urban areas, like the resident killer whale pods that show up in Puget Sound near Seattle, their toxin-laced blubber may be a detriment to their health and reproductive fitness during lean times.

Killer whales are built on the same body plan as many of their smaller relatives, oceanic dolphins such as bottlenose, spinner, or common dolphins, except with fewer and larger teeth, and scaled up several times. When we chart the ratio of brain size to body size—a metric called the encephalization quotient, or

EQ—we have a way of quantifying the fact that dolphins are indeed very brainy. While baleen whales and river dolphins plot closer to primates, oceanic dolphins—including killer whales—plot higher than every other mammal except us, slotting in second behind humans, but ahead of chimpanzees. For most of the past ten million years, a dolphin was the brainiest creature on the planet. The EQ values for our own lineage bumped up beyond the dolphin threshold only in the past few hundred thousand years.

In the rare opportunities when it's possible to preserve their brains after death—brain tissue liquefies extremely quickly unless it's stabilized with preservatives—we can see a highly corrugated exterior surface across two bulbous hemispheres, much like our own. That shape and texture of the cerebral cortex is very different from that of, say, a deer, sheep, or cow, which are among the nearest relatives to cetaceans. It's difficult enough to infer function from form in human brains—after all, figuring out how the architecture of 100 billion neurons writes symphonies is one of the great unresolved questions in modern neuroscience. But the deep gyri and sulci on dolphins, which are similar to our own, tell us that there is some anatomical underpinning to the behaviors that appear so sophisticated and intelligent to us.

If we share similarly structured brains with killer whales, are there behaviors beyond the facts of their ecology—such as cooperative hunting or family structures lasting decades—that we share as well? For behavioral biologists, mirror self-recognition

is a coarse way to determine if the viewer has a sense of awareness, which in turn suggests the ability to understand that the
animal on the other side of the mirror isn't a random part of its
world. The list of mammals that can look into a mirror and
recognize their own faces is very short: us, great apes, Asian
elephants, and possibly two kinds of cetaceans—bottlenose dolphins and, maybe, killer whales. The test is straightforward
in concept—place a mark on an animal and then determine
whether it examines the mark in the mirror—but it is extremely
tricky to execute in a controlled, experimental setting, especially for an aquatic mammal without opposable thumbs. Bottlenose dolphins in captivity have afforded the best opportunities,
whereas experiments with captive killer whales have been more
tricky. Regardless of logistics, the best data available show that
mirror self-recognition in cetaceans is real. These whales are
quite aware that they have lives circumscribed by plain concrete
walls, instead of a world as complex and dynamic as an ocean.
What we do with that information depends on our response as
a social species—our laws, ethics, and economics all playing
into the decisions we make about keeping large oceanic predators captive for our own entertainment. If EQ is broadly reflec-
tive of some kind of cultural capacity, then it's a fair supposition
that cetacean societies have been a part of this planet for at least
ten to one hundred times longer than our own. How old are
some of their traditions and behaviors? What cultural innovations did they have that went extinct? We can read only so much
from bones; maybe we'll figure out how to scratch away at those
other questions.

One way that we can tackle some of these issues is to examine the complexity of their acoustic lives. The iconic bellowing and rubbery chirps of whalesong are just some of many kinds of vocalizations by baleen whales; some strange and unusual sounds heard in the oceans still haven't been entirely tied to specific species. (Only recently have researchers connected a bizarre underwater *boing* noise heard for decades to minke whales.) Toothed whales, such as dolphins, beaked whales, and sperm whales, vocalize at low frequencies too but also use high-frequency or ultrasonic clicks and pings emitted from their biosonar apparatus to navigate and signal to others around them. It's a bit like an acoustic analog of running around at night using flashlights with your friends—by light alone you can see where they're going and what they're spotlighting and infer something about their movement. The click trains of echolocating toothed whales are complex and so variable in frequency, power, and organization that the mathematics of information theory seems to be one of the few tractable ways to describe their information content.

The fact that whales seem to be having conversations without us hasn't been lost on scientists eager to bridge the divide. This divide, however, is many times greater than that between us and chimpanzees and other great apes; at least we share common body form, ecology, and communication modes. With whales it has been an immense challenge to interpret the important threads of information that they share among themselves. A click train from a sperm whale a mile deep may convey thoughts

about where lunch is located or the meaning of the universe—
we cannot tell or know the difference because the meaning of
their language is lost without knowing the context. All attempts
to decipher some kind of language out of the jabber of squeaks,
chirps, and rapid taps have failed because whales live in a world
that is so foreign to us.

There is one last piece to the still-unsolved puzzle of whale intelligence, and it fits somewhere between behavior and cognition:
whales possess culture. Let's put aside art and material culture
for a moment and focus on the broader definition of culture: any
kind of information stored outside an animal's DNA that can be
transmitted across individuals or generations. That type of definition fits what many species of whales do, as much as chimpanzees and elephants, among other animals besides us.

Unsurprisingly, the best available evidence for culture in
whales comes from their acoustic behavior—that content-rich
but enigmatic repertoire of clicks in echolocating whales' calls.
Fish-eating and mammal-eating killer whale ecotypes each possess separate and distinctive acoustic repertoires, cohesive across
several matrilineal groups numbering in the hundreds and ranging over hundreds of miles. Sperm whales, which have distinct
acoustic clans, take these linkages of cetacean societies and expand them to the scale of ocean basins, stretching over thousands of individuals and over thousands of miles; they may
represent the most far-flung cooperative groups aside from us.
Sperm whale clans are made up of multiple smaller, long-term
social units, akin to pods among orcas. Each clan is tied to

multiple matrilines that hang together in multigenerational societies of sperm whale grandmothers, mothers, and daughters, all learning from one another how to hunt squid, raise calves, babysit, and defend against their only predators—killer whales. (Adult males live comparatively solitary lives after leaving their natal social units in their teens, and are sporadic members of sperm whale societies.) Each sperm whale unit shares the same distinctive acoustic dialect, and these dialects form a common basis for interacting at sea. Imagine, for a moment, what whaling might have done to these societies. Losing an entire clan of sperm whales, as rapacious as whaling was, may have meant losing an entire dialect and all of its distinct cultural traditions with it.

Given what we know about past effects of humans on whales, which of them now stand the best chance of success in the Anthropocene? Several imperatives give us clues about which whales are likely to be among the winners. First, be just the right size. Almost all of the largest whale species today, including blue whales and right whales, are navigating an increasingly urbanized ocean habitat while still struggling to recover from the centuries-long legacy of industrial whaling. Blue, fin, bowhead, and right whales all rank among the largest animals ever to have evolved on the planet, with body weights exceeding eighty tons, reaching well beyond one hundred tons for the largest individuals. While these sizes likely evolved out of trade-offs for feeding efficiency and long-range migration, their extreme size today puts them at risk of entanglement in fishing gear and trauma from ship strikes.

At the other end, many of the smallest whale species, such as the vaquita and finless porpoises, are in jeopardy of extinction because their body size evolved as a consequence of their geographic isolation. In their case, changes to a single river or embayment can imperil the entirety of their lineage. Most whale species fall in size ranges between those extremes; their fates are instead tied to other factors, such as diet and geographic range.

Second, don't be a picky eater. In an evolutionary view, hyperspecialization allows species to exploit particular lifestyles, habitats, or prey resources—consider parasites dependent on a single host species, or hummingbirds that feed on the nectar of a specific flower. Should any component of that relationship change, the hyperspecialist may be at an evolutionary dead end. Whales specializing in one specific prey type, even if it is particularly abundant—such as blue whales dependent on krill or killer whale ecotypes eating exclusively salmon—may have little flexibility to adapt to sudden changes in the environment, especially if changes in ocean chemistry and temperature affect their food security. Whales that can feed on a variety of prey species, such as humpback or gray whales, will have more flexibility than whales that feed on prey available only in some places or at certain times—humpbacks and gray whales also happen to be just the right size, falling in a Goldilocks size category for baleen whales that's not too big and not too small. The rich diversity and enormous abundance of cephalopods in the open ocean comprise a large portion of the diet of toothed whales such as sperm whales, beaked whales, and oceanic dolphins; cephalopods show no signs of going away in the Anthropocene, providing at least some measure of food security for these species.

Third, stay global. Having a global footprint provides an insurance policy against local or regional calamity, maximizes gene flow, and broadens access to a variety of prey. Sperm whales, killer whales, and humpback whales are all found across the world. Even if some regional populations are distinctive—killer whales, for example, have genetically restricted ecotypes with specific diets; some humpback populations in the Arabian Sea do not migrate at all—the rest of the lineage is broadly enough distributed around the globe to provide the greatest possible hedge against extinction.

Fourth, culture helps. Killer whales, humpbacks, and sperm whales all possess a kind of culture that is woven into their behavior and social structure and provides resilience against unexpected changes as well as the basic ability to innovate in a changing environment—culture is adaptive, after all. Anthropocene Earth is certainly changing rapidly, and the hazards for cetaceans are myriad and complex across geography and time. Whale species such as river dolphins and the vaquita don't appear to show the same cultural traits that we find in other cetaceans, although we still know so little about the minds of whales that it is presumptuous to apply a harsh rubric on this factor.

Last, nothing's perfect. The factors that lead to success or peril in the Anthropocene don't conveniently apply in a clean way across all whale species alive today. For example, blue whales are globally distributed, but they push the maximum size limit for rorqual whales and have a very specialized diet; they were also the target of sustained pressure eliminating 99 percent of their biological diversity. They are almost paradoxically awesome: what makes them marvels of efficiency has also made them vulnerable to extinction. Or take killer whales again: they range across the entire globe, have a broad dietary range (as a species), and live in large groups with strong social units (and culture), all of which provides a vouchsafe against the unpredictable and ongoing changes to their environment. However, their position at the apex of the food chain also makes them susceptible to persistent pollutants that biomagnify and concentrate in their bodies. Killer whales, like blue whales, live at the mercy and curse of human civilization.

When I think about the killer whale skeleton hanging in Sitka, I think about how killer whales remain a story of evolution right before our eyes. As a species, their prey diversity is quite large, but the different kinds of ecotypes, with their highly specific diets (and cultures), means that the genetic divergences among these ecologically distinct populations represent the first steps of true lineage splitting leading to new species of killer whales, despite the fact that they live literally side by side. There seems to be no going back for the mammal, salmon, and shark eaters.

Like us, killer whales have big brains and lead complex lives in complex environments—traits that put some of their lineages at risk, while simultaneously enabling other lineages to continue to evolve in new ways. Maybe the present day for killer whales isn't too different from our own recent evolutionary past, when many big-brained hominids competed for Pleistocene resources. There are, fortunately, many mysteries left for us to uncover as we try to navigate the open waters between our two cultures, cetacean and human.

17.

WHALEBONE JUNCTION

The low light of a fall afternoon stretched the shadows as we made our way across the Outer Banks of North Carolina. Out the car window I scanned the dunes and the windblown oak scrubland. My wife and son gazed out, lost in thought, while the toddler slept, her cheeks still and chest lifting in quiet pace. I turned off the state road, easing the car to a stop that interrupted everyone's daydreaming. "Okay, we're here," I announced.

"This isn't a park," my son declared after glancing out the window. "You said that there would be a park. With swings." My wife winced as she looked at me. "I know," I said. "I'll tell you what: I just need a few minutes. Then we'll go find that park. Or climb to the top of the Bodie Island lighthouse. We haven't done that one yet," I offered. We tried to visit every lighthouse that we could along the coast. My son is fearless of heights, in the way that children can sometimes be so different from their parents.

He moaned. "I'll be fast," I said, not sure that I could back that up. "But we're not carrying specimens for the National Museum this time, so you don't have to stay and guard them. You can come in with me if you want." I smiled at my own little joke, which no one else found funny. The things that you ask your family to do when you're a paleontologist.

I walked out to the small hut of a ranger station, where a brightly painted sign read "Whalebone Junction." This spot, at the northern gate to the Cape Hatteras National Seashore, had earned its name from a whale skeleton that had once been mounted in front of a gas station, decades ago. It stood as a waypoint between the road heading north to Kitty Hawk and a turn east, across a bridge to the mainland.

Today both the skeleton and the gas station are long gone. The name, of course, piqued my interest, but for a reason that had more to do with Deep Time than the last few decades: it's not always clear if whale bones on a beach in this part of the world belong to a living species. It's possible they belong to completely extinct ones, occasionally roughed out of fossil-rich rocks in rock layers underwater, just yards away from the beach. These bones can be relics from the ice ages or many more millions of years further back in geologic time. You just never know. I thought that chatting up a local might yield more than I could scrounge on the Internet. Inside, the volunteer offered little beyond vague recollections and an Internet search or two on the computer at hand. Bemused, I thanked him and returned to the car.

"See?" I boasted when I returned. My family was unimpressed. "Lucy's awake," my son intoned matter-of-factly. "I think it might be ice-cream time before we go to the lighthouse," my wife announced, looking to the back of the car. "Sounds about right to me too," I said.

I thought of the one historical black-and-white image that I had seen of the whale at Whalebone Junction, maybe from the 1950s. It showed several children leaning against a sad excuse for a skeletal mount: a broken skull tilted against a string of vertebrae, with a jawbone slid underneath in a comically incorrect pose. The skeleton probably belonged to a humpback whale, a sei whale, or another midsize species of rorqual. But the reason why I had followed this thread more than the usual loose threads was the hope that it might have been a gray whale—a species closely related to rorquals but lacking the flexible throat pouch and titanic size classes. The Outer Banks is the wrong place to chase living gray whales today because there are none to be found. It is, however, a place to chase their past—and their future.

Today gray whales are among the most abundant whale species in the North Pacific Ocean. True to their name, their skin is the color of gunmetal, underlying a patchwork of mottled white and yellow splotches. Their snout is more beaklike than that of any baleen whale, and the rest of their body outline a bit unremarkable, except for a knuckled tailstock. They swim slowly and, unlike most baleen whales, they mostly hew close to the shoreline, where sharp-eyed whale watchers can spot them from atop seaside overlooks—or certainly their uniquely puffy, heart-shaped blows. The most recent surveys place their total numbers at some twenty thousand, unevenly distributed across two populations along the western and eastern coasts of the North Pacific basin, stretching from the Korean peninsula to Baja California. Nearly every gray whale alive today belongs to the so-called California or eastern population; western gray whales, which live mostly in the Sea of Okhotsk off Russia, number perhaps no more than one hundred and remain poorly studied.

As a consequence of their lifestyle and abundance, gray whales are also iconic whales of the North Pacific—like cave paintings for the modern world, their life-size likenesses grace murals on parking garages and downtown business centers from San Diego to British Columbia. Their lore has a lot to do with the fact that the eastern population undertakes one of the longest migrations of any mammal on the planet: after overwintering in lagoons along the Pacific side of Baja California, they migrate with new-born calves to feed on the shallow seafloors of the Bering and Chukchi seas during the bountiful summer. Then, as summer turns to fall, they return from Alaska to Baja, completing an approximately ten-thousand-mile round-trip. Small towns from Sitka to Monterey time festivals to celebrate their passage within eye- and earshot, with the requisite T-shirts and tchotchkes.

Gray whales seek the shallow seafloors off Alaska because these habitats are hugely productive, harboring an incredible abundance of crustaceans. During the summer months, increased sunlight jump-starts the amount of energy entering this marine ecosystem, which produces enough biomass to support millions of seabirds, walruses, halibut, and whales. Gray whales are unusual among baleen whales in having an especially broad diet, and they have a similarly flexible feeding style: like a vacuum, they can suck and filter carpets of soft seafloor loaded with invertebrate prey; or they can chase and gulp swarms, even schools of fish, in the water column. The feeding trails left on the seafloor by gray whales are wide, long, and in high enough density to completely restructure it, which has led some scientists to describe gray whales as ecosystem engineers—by stirring up sediment and leaving a wake of suspended organisms, they create the food webs for other, smaller animals in the Bering and Chukchi seas to consume.

Their coastal preference, however, also nearly became their undoing. Charles M. Scammon, a midnineteenth-century whaling captain with the keen eye of a naturalist, published the first observations of living gray whales when his crews discovered their breeding grounds in Baja California. Within decades, gray whale numbers plummeted as they became easy targets for whalers in the North Pacific. By the end of the nineteenth century, it wasn't even clear if any gray whales still survived. As gray whales continued to be killed into the twentieth century, the IWC eventually implemented a commercial whaling ban on this species in 1946. Over the course of the ensuing decades, well through the peak of Southern Ocean whaling and illegal Soviet whaling in the North Pacific, gray whales managed to recover. In 1994 NOAA officially delisted gray whales from their endangered status under the U.S. Endangered Species Act, recognizing their recovery from the brink of extinction in less than a hundred years.

Since their delisting, gray whales' population has mostly stayed constant, and they are frequently touted as a conservation success story—in marked contrast to other species of large baleen whales, which were decimated over the same time period. Protection from whaling is likely what gave the species breathing room, letting their biology do the rest. But really understanding gray whales—and their potential future—requires stepping much further back in time and looking to horizons beyond the North Pacific.

My colleague Scott Noakes is a scientific diver by profession, and what I do on land he does underwater, many miles offshore: he collects bones, but he does it from the seafloor. He uses a hose instead of a rock pick to blow sediment away, and dive flippers instead of boots. We both share limitations of access, but where timing is concerned, I might worry about losing fossils to the tides; he has to worry about running out of oxygen.

Bones belonging to large, extinct mammals are not uncommon finds in the waterways and underwater coastlines of the southeastern United States. But instead of finding fragments of ice age mammoths or camels off the coast of Georgia, Scott found whale bones. Each episode of sea-level rise and fall over the past few million years has created unique reefs along the coast, today located in places that are too deep and too cold to support corals. Clearly whales once lived—and died—nearby. Scott and his colleagues found jawbones belonging to not one but two different individual gray whales, both young animals, based on their length. In all likelihood this part of the world was once part of the migration path for Atlantic gray whales, if not very close to their calving grounds during the ice ages.

Advanced dating techniques placed the two sets of bones at about 41,000 and 48,000 years old, well into the last glacial interval. Scott's gray whale bones are the oldest of any found over the past few decades along eastern shores of the United States, from Myrtle Beach and the Outer Banks in the Carolinas through New Jersey to Long Island. Gray whale bones have been discovered as far away as England and Sweden, showing a range of occupation across the North Atlantic not unlike that of their counterparts across the North Pacific today. Radiocarbon ages extracted from some of these bones—including specimens stored at the Smithsonian—provide a clear chronology for gray whales since the last glacial episode of the ice ages up until about 450 years ago. This time frame ends well before the nineteenth-century heyday of American whaling. So when and why did Atlantic gray whales go extinct? Until we get more fossils or more recent bones with DNA, we won't know more about this ghost lineage of gray whales.

If gray whales were once in the Atlantic and today live only in the Pacific, how did they ever cross ocean basins? Changes in sea level would have provided few portals across oceans. Even at times of sea-level highs, the terrain of Panama is still too high to cross. A trip around Cape Horn, off South America, would require a dispersal equivalent to the circumference of the entire planet, over 26,000 miles. That distance is perhaps not outside the realm of possibility for a species of whale that migrates half that far on an annual basis, but there is a much better reason to suspect that the answer lies in the Northwest Passage. It's not just because the Canadian Arctic would have afforded an open passage during warm times and high sea level. Rather, it's because gray whales are already probably using the Northwest Passage to get back into the Atlantic right now.

I remember seeing the first reports of living gray whales outside the Pacific surface on an Internet message board. A few years ago, researchers in Israel photographed a whale that they

had never seen before. The only identification they could hazard
was gray whale: a pinched, heart-shaped tail fluke, mottled gray
body coloring, and especially a knuckled tailstock. This same
individual whale—identified by the white splotches on its tail
fluke—then turned up off the coast of Spain several months
later. And then came similar reports, with photographs, of another gray whale spotted off the coast of Namibia, well south of
the equator.

Researchers concluded that the most likely pathway for these
errant gray whales was through the Northwest Passage, probably during a nearly ice-free summer. The straight-line distance
of this migration from Baja to Israel via the Arctic adds up to
nearly 24,000 miles, which would be an astonishing figure for
a single lost whale. But we know that this figure is about the
same as the distance from Baja to Namibia. At the moment,
these two examples might just be anecdotes—or harbingers of
systemwide shifts in the Arctic.

The ongoing and future changes to the Arctic essentially
mean a large-scale transition of its ocean from an ice-covered
one to an open, or pelagic, one. In other words, the Arctic
Ocean will start to look more like oceans at other latitudes, but
with polar seasonal settings including long summer daylight
hours. The sunlight, unimpeded by ice, will boost trophic productivity, increasing the prey resources on the seafloor and in
the water column, which is ideal habitat for gray whales, among
other species. Over the course of a few whale generations, movement through this passage may be more than mere dispersal to
a newly available habitat—it might become a future migration
route for whale species that already migrate thousands of miles,

such as humpback, sei, minke, and even right whales. If the Northwest Passage was indeed their dispersal route, the projected ice-free summers in a twenty-first-century Arctic bode well for more Atlantic invaders. The future of gray whales may be closer than we think.

After stopping for ice cream, we pulled up to the lighthouse parking lot a few minutes' drive away from Whalebone Junction. An avalanche of toys, books, and foodstuff spilled out of the car as the kids climbed out and made a beeline for the lighthouse. A square, two-story house that formed the keeper's quarters stood directly in front of the lighthouse, a hundred-fifty-foot-tall brick building built in 1872, striped in black and white.

I like lighthouses: their picturesque seaside settings, the views they offer, their history. Lighthouses were critical pieces of coastal infrastructure and national security in the nineteenth century. In 1883 one of my predecessors at the Smithsonian, Frederick William True, proposed that lighthouse keepers record whale strandings. It would take nearly one hundred years for NOAA to establish the first stranding networks, which now capture the what, when, and where of any whale stranding reported in the United States.

Back down on the ground, our children ran on the brick path from the lighthouse to the old lightkeeper's house nearby. My son outpaced his sister in the spontaneous race, opening a gap. My wife, walking close alongside me, grabbed my hand and squeezed my fingers. Lighthouse keeping is no more, replaced by automation; car trips to this place also might soon be a thing of the past, depending on how fast sea levels rise. Light-houses along the coast have already been moved against the receding shorelines. I wondered how many more times light-houses would be moved again—and what my own children might witness. Trying to understand the transformations that have happened in the planet's past captivates me as a scientist; as a parent, I wondered how much I could do to prepare my children for their future.

The next day we sat high on the beach at Kill Devil Hills, just north of Whalebone Junction. The kids dug endlessly in the surf, poked sodden flotsam, and occasionally faked a run toward the waves. I found myself distracted, staring at the fine line of the horizon between slate-green water and azure sky. Gray whale bones at the Smithsonian, in the collections that I oversee, were collected just south and north along this same stretch of coast-line, only miles away in Nags Head and Corolla. The bones are so worn and roughed up that they can hardly be called beauti-ful, but they demarcate in a real way past worlds that perhaps no one saw. Gray whales were probably critical members of ancient Atlantic food webs, as much as they are in the Pacific today. How far and where did Atlantic gray whales migrate? They sur-vived many episodes of glacial ice sheets and sea-level drops that changed their habitat—how and why did they disappear from the Atlantic? Did they leave any genetic traces in today's Pacific gray whales? All of these questions fundamentally stem from a historical perspective on whales that is contiguous with what we see today, not just a pale, creased vignette of the past.

Past whale worlds aren't really dead; they're not even past, to paraphrase Faulkner. Thinking about those bones and those questions made me think that on a fall day in a few decades, it might not seem strange to see gray whales spy hopping along the Atlantic coastline once again. Maybe they'll do so alongside North Atlantic right whales, which narrowly averted extinction, and ever-present bottlenose dolphins. I imagine these gray whales, newly returned to the Atlantic, carrying biologging tags and accompanied by drones—technologies created as an extension of our senses to help us spy more effectively on their hidden lives. And maybe the view will be from a boat, not a sandhill, if we can't figure out how to stop the rise of the seas, should the great glacial ice far north and south all melt. Places like the Outer Banks would slip under the waves, returning any still-undiscovered whale bones of past gray whales back to the water. I don't doubt that there will still be scientists eager to dive after them, or chase after the whales navigating the seas of a new world.

EPILOGUE

The warm water of the Chesapeake Bay lapped over my feet as I walked behind my son, who scoured pebbles in the surf. I kept an eye on him as my colleague Dave Bohaska walked beside me. I knew that we stood a good chance of finding a few fossil shark teeth and some nice shells on this western side of the bay. The tall Calvert Cliffs, which line this part of the Chesapeake's shoreline in Maryland, have produced hundreds of thousands of Miocene-age fossils, including many skulls and skeletons of fossil whales. All of my predecessors at the Smithsonian—people like Frederick True and Remington Kellogg—either knew about or collected fossil marine mammals here.

It was supposed to be a leisurely Sunday, and we held no high hopes for a significant find. I hadn't brought any tools, and I mainly wanted to unplug and spend time outside with my son; maybe fossils would inspire him as they did me, but I wasn't going to push it. Dave lives nearby, in the community of Scientists Cliffs, and he was happy to take us on a neighborhood walk. If nothing else, the community's name alone is sufficiently auspicious for casual fossil finding.

Abruptly my son paused next to a block of sediment that had slid down to the beach. "Dad—what's this?" he called. I walked

up and knelt down for a closer look. Dave and I are both familiar enough with these kinds of anatomical riddles to immediately recognize that it was the snout of a whale. The rest of the
skull was either still trapped high in the cliff face from which
this block had tumbled or lost to the bay. On closer inspection,
the right and left snout bones were strongly skewed, near where
the face housed a biosonar apparatus. I thought that it probably
belonged to *Orycterocetus*, a fossil sperm whale that Kellogg
worked on extensively, especially from these rocks. "Well, gee,"
Dave said, smiling. "Anders," I said proudly, "you just found a
fossil whale." My son, then four years old, wasn't quite sure
what to make of the situation.

I stood up and exhaled. "Dammit," I said softly. Deciding
against bringing excavation gear guarantees that you'll need it.
No hammer, no plaster, no pickax. Dave and I traded glances.
"It's at tide level, fair game for collecting," Dave remarked, implying that we were clear, legally, to do so for the Smithsonian.
I reached into my day pack and pulled out a small knife; Dave
had one too. We were losing daylight and didn't have time to
hike back to Dave's house for more tools. The tide was coming
in; who knew if the snout would still be there the next day. I
looked at our pocketknives and considered the muddy, water-
logged sediment. "All right. Let's do it," I said, and we started
digging.

Sometimes finds of happenstance are curiosities; sometimes
they are important scientific discoveries and should be collected.
I thought the snout was probably the latter, well worth the

trouble. Sperm whales have one of the deepest evolutionary histories of any living group of whales, stretching from the iconic sperm whale of *Moby-Dick* back to fossils from the Oligocene, more than 23 million years ago. While sperm whales in the fossil record may differ in details of tooth size and subtle sutures of the skull, they all share a bowl-like concavity on their face housing the junk and case—special terms for their biosonar organs, owing to nineteenth-century whalers who sought their fine and lucrative oil. Fossil sperm whales can be found all around the world, represented mainly by large, handsome teeth, sometimes the weight and size of a wine bottle. We had found them at Cerro Ballena; today's sperm whales' polished teeth are the stuff of scrimshaw and lore. Ivory anvils of ten thousand squid meals.

We scraped away sediment to guess the dimensions, in keeping with an *Orycterocetus* snout. We would need to carry it up several staircases to where we'd parked. The snout had no teeth with it, unfortunately, just deep grooves on the palate, where they once were held in place by ligaments. Still, it was more than I had ever found at the cliffs. I took off my T-shirt to make a sling for the block.

As we trenched out the snout, I occasionally shot a glance at my son, who chucked handfuls of rock into the tide, lacking anything more substantive to do. I thought about how at one point about fifteen million years ago, this snout belonged to a living, breathing sperm whale. This long-extinct individual whale had a life history, a diet, and it belonged to a vanished ecosystem; I wondered about the owner of this snout, and thought about questions that paleontologists would have trouble answering: Was it a mother? Was it part of a society, with culture? Would this Miocene whale have been able to communicate with its descendants alive now? Did it sleep vertically, in a floating forest of leviathans, just below the surface, the way sperm whales sometimes do today?

We individuate whales as a way to know to them, to chip away at their mystery, whether it's a fragment of skull, the unique squiggle of an acoustic click's waveform, or a distinctive mottling on a tail fluke. We give individual whales familiar names or alphanumerics to differentiate them from their kin, as if that somehow circumscribes their enigma, easing our inquiry into their history and lives. Individual whales matter because they sometimes carry the superlatives that scientists bestow; individuals are, after all, tied to the records of the largest, the deepest, the first, and, sometimes, the last.

But this is how science happens: collect these individuals together, under a specific question, or a line of inquiry, and you can build a broader picture that begins to illuminate the inaccessibility of whales. I interrogate individuals when I pick up a skull or a bone and ask, "Who are you?" That's my starting point for discovery. And I know that chain of questioning is not unique to me—it's the same question asked of the plug of tissue stuck in a biopsy dart, or through the observation panel of an aquarium, gazing at a captive dolphin. Who are they? That's ultimately what we all want to know about whales.

Later that evening, my son in bed, I headed to the National Museum of Natural History in downtown Washington, D.C., to haul the block containing the snout to my laboratory. I rolled a cart over to my parked car in the dark and unloaded the skull bones, still wrapped in a soaked T-shirt. Earlier that day, the fossil had been an anonymous hunk of bone stuck in marl. Now it would be prepared, cleaned, measured, studied, and preserved for generations to come. It would receive its own catalog number—USNM 559329—with my son recorded in the database as its collector. It would join the ranks of thousands of other fossil specimens like it at the Smithsonian, each one with a story to tell. Those stories are discoveries waiting to be made—some mundane, others that could rewrite textbooks—and that's what makes science so much fun. Teasing out the facts from the unknown depends on asking the right question; the answer might tell you a lot about the past, present, and future of whales.

ACKNOWLEDGMENTS

I had been thinking about writing this book for many years. It started with vignettes that I had collected while doing science all around the world. The sketches and stories gelled over the course of swaying subway commutes, staring out airplane windows, and quiet moments in the field—aboard ships, at rock outcrops, and along windswept shores. These snapshots represent many threads from investigating the past and present lives of whales, something that makes the stories in this book about the people who contribute to science as much as it is about whales. Foremost, I am especially grateful for the kindness and hospitality of colleagues in Santiago, Caldera, Lima, Panama City, San Diego, Los Angeles, Bakersfield, Pacific Grove, Santa Cruz, Berkeley, Newport, Seattle, Victoria, Vancouver, Sitka, Ashoro, Melbourne, Wellington, Dunedin, Folly Beach, Scientists Cliffs, and Reykjavik.

I am profoundly thankful for the trust and camaraderie of Bridget Matzie, who along with Esmond Harmsworth saw the bigger picture in the first fragments of undigested text. This book would have stayed in vignette form if it weren't for Emily Wunderlich's dedication and sharp editorial pen. I am so grateful for her ability to spot a narrative from rough shards of half stories and hear the precise meaning of an idea obscured by poor form, annoyance, or exhaustion. Shannon O'Neill was indispensable as a wordsmith, whisperer, and therapist. In an early stage, Melanie Tortoroli's exhortations and guidance set me on the right course. I knew that this book found the right home with all of the staff at Viking. I was constantly reassured by their professionalism and attentive care, especially from Hilary Roberts and Cassandra Garruzzo. I am also grateful for the persistence and support from key members of the Smithsonian side of this book, including Carol LeBlanc, Brigid Ferraro, Sue Perez-Jackson, Carolyn Gleason, and Eryn Starun.

Alex Boersma crafted every illustration in this book. Each one is an original piece made in linocut or scratchboard, specific to the stories in this book and generated out of a common way of looking at the world. Alex was uniquely suited for the task, not merely as an artist and a scientist—and she is definitely both—but also because of her own firsthand knowledge of all things cetacean, by pneumatic chisel, scalpel blade, and tagging pole. Thank you for coming along on the journey.

From the outset, there were people who intuitively understood why I would write a book like this one. You should be so lucky to have the kind of support that I received from Ari Friedlaender, Jeremy Goldbogen, Carolina Gutstein, Kris Helgen, Katie Kuker, Gene Kwon, Holly Little, Megan McKenna, Jessica Meir, Aaron O'Dea, Jim Parham, Kris Perta, Aviva Rosenthal, Caroline Stromberg, Lesley Thorne, Jann Vendetti, Steve Weyer, and Greg Wilson. You each restore meaning to the most commodified word of our time: friend.

You also want your book to be read or, at least, be readable. I'm grateful for the encouragement and thoughtful comments, at various stages, from Lauren Appelbaum, Kay Behrensmeyer, Matt Carrano, Daniel Epstein, Doug Erwin, Molly Fannon, Ari Friedlaender, Jeremy Goldbogen, Gene Hunt, Sarah Hurtt, Randy Irmis, Dave Lindberg, Kevin Padian, Jim Parham, Aviva Rosenthal, and Bob Shadwick. There is also nothing like the unvarnished truth from family. Sue Hunter, Tom Hunter, Mitch Hunter, Ben Pyenson, Catharine Pyenson, and Lewis Pyenson read through many drafts and never hesitated to tell me where I went wrong.

In the early stages, Neil Shubin, Scott Sampson, Tim Flannery, Ed Yong, and Sean B. Carroll offered requisite bookwriting advice. Passages in this book about the vaquita also first appeared in an article for *Smithsonian* magazine in 2017. At key moments, I benefited from the support of Wayne Clough, Sarah Goforth, Nancy Knowlton, John Kress, Hillary Rosner, Sabrina Sholts, and Lydia Pyne. I especially want to thank Jan and John Straley in Sitka and John and Terry Miller in San Pedro—our conversations, over a meal, were important to me in ways that they didn't know.

I thank all of the dedicated staff who contribute to the many-sided mission of the Smithsonian Institution, which is a constellation of museums, research centers, and field stations, among other entities. I would like to especially highlight the support that I've received for my own research program from the Smithsonian Tropical Research Institute, the Office of International Relations, Smithsonian Libraries (including all of the divisional libraries), Smithsonian Enterprises, the Office of General Counsel, the Office of Public Affairs, and the Digitization Program Office. Closer to my own world, I thank the offices of the Director, Communications, Development, the Associate Director of

Science at the National Museum of Natural History, and the terrific staff in my own Department of Paleobiology.

Aside from those mentioned in the book, I leaned on the expertise of many people in the process of writing this book. I thank them for their time and hold them blameless for my own mistakes: Renee Albertson, Lars Bejder, Gunnar Bergmann, Trevor Branch, Paul Brodie, Bob Brownell, Graham Burnett, Ellen Chenoweth, Fredrik Christiansen, Phil Clapham, Erich Fitzgerald, Ewan Fordyce, Shane Gero, Stephen Godfrey, Dalli Halldórsson, Yulia Ivashchenko, Jen Jackson, Dave Johnston, Bob Jones, Igor Krupnik, Kristin Laidre, Halldór Lárusson, Jacobus Le Roux, Kristjan Loftsson, Lori Marino, Chris Marshall, Rocky McGowen, Jim Mead, Dick Norris, Aaron O'Dea, Droplaug Ólafsdóttir, Daniel Palacios, Charley Potter, Stephen Raverty, David Rubilar Rogers, Angie Sremba, Gabor Szathmary, Hans Thewissen, Kirsten Thompson, Gisli Víkingsson, Wayne Vogl, and Alex Werth.

While writing this book, I also leaned heavily on all of the members, past and present, in my research group. I owe a debt of thanks to Carlos Peredo, Matt Leslie, Aly Fleming, Ani Valenzuela-Toro, Jenell Larsen, Jorge Velez-Juarbe, Fri Engel, Holly Little, Neil Kelley, Maya Yamato, Matt McCurry, Dave Bohaska, and Carrie Carter. I wrote a large portion of this book during a winter in Alaska as a visiting scientist at the Sitka Sound Science Center through their Scientist-in-Residence Fellowship program. I warmly thank Lisa Busch, Tori O'Connell, Lauren Bell, Dane McFadden, the Straleys, and Davey Lubin for all of their hospitality.

I am, beyond all else, indebted to my wife, Emily Hunter Pyenson, who kept my world together during the prolonged hellfire of book writing, reading countless drafts, clarifying ideas with incisive thought, and reminding me about the big picture. I hope that our children might read this book one day to understand a bit more about why their father cares about bones, skeletons, and mysteries pulled from the sea.

A FAMILY TREE OF WHALES

Some quick words about terminology: Scientists recognize over eighty species of whales alive today, and they all could equally be called cetaceans. The word "cetacean" comes from the Latin word *cetus*, and it was codified by Carl Linnaeus in the eighteenth century to describe the collective taxonomic group comprising all living whales. Cetaceans fall into two great branches of a family tree: baleen whales, also called mysticetes; and toothed whales, or odontocetes. While these two groups have evolved in separate ways for tens of millions of years, with one evolving many different ways to filter feed and the other one using echolocation, they are each other's closest relatives—what evolutionary biologists call sister groups. There are hundreds of fossil whales inside these two main houses of whales, with some more closely related to mysticetes and others more closely related to odontocetes. They descend from a common ancestor that lived in the oceans more than 35 million years ago.

Early branches in the family tree of whale evolutionary history are represented by a variety of fossil species. Some of these were four-legged and lived much of their lives on land; others seem to fall somewhere between those land-bound whales and today's fully aquatic whales. In this book, I call these extinct lineages "early whales," although some scientists have called them archaeocetes. Below is a general diagram outlining the evolutionary relationships—as best we know them—for these broad groups.

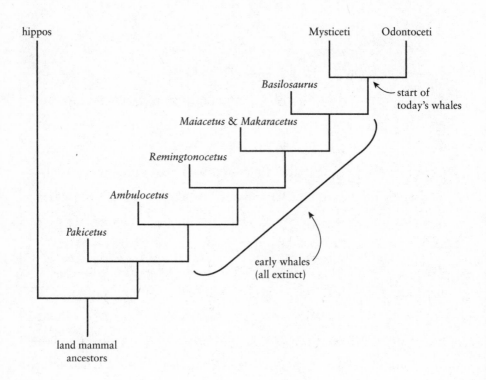

Throughout the book, I use common names for most whale species alive today, and scientific names for most fossil whale species, usually only the first (or generic) name of their two-part Latinized binomial (genus and species). Below, I've provided a framework for understanding the general taxonomic groups for each whale species, living or extinct, mentioned in this book. Not every higher taxonomic category has a convenient nonscientific shorthand; I've alphabetized those that do, where they exist, and otherwise alphabetized by scientific name for groups that lack them. Also, as a caveat, this list is not meant to imply evolutionary relationships among the higher groupings of whales. The family tree of whales remains a work in progress.

Taxonomic Group		Common Name	Scientific Name
Mysticeti (baleen whales)		Pygmy right whales Gray whales	*Caperea marginata* *Eschrichtius robustus*
	Right whales	Bowhead whales Southern right whales North Atlantic right whales North Pacific right whales	*Balaena mysticetus* *Eubalaena australis* *Eubalaena glacialis* *Eubalaena japonica*
	Rorqual whales	Minke whales Antarctic minke whales Sei whales Blue whales Fin whales Humpback whales	*Balaenoptera acutorostrata* *Balaenoptera bonaerensis* *Balaenoptera borealis* *Balaenoptera musculus* *Balaenoptera physalus* *Megaptera novaeangliae*
Odontoceti (toothed whales)		Belugas Narwhals Walrus whales	*Delphinapterus leucas* *Monodon monoceros* *Odobenocetops peruvianus*
	Beaked whales	Cuvier's beaked whales	*Ziphius cavirostris*
	Oceanic dolphins	Common dolphins Killer whales or orca Spinner dolphins Bottlenose dolphins	*Delphinus delphis* *Orcinus orca* *Stenella longirostris* *Tursiops truncatus*
	Porpoises	Finless porpoises Vaquita	*Neophocaena asiaeorientalis* *Phocoena sinus*
	River dolphins	Amazon river dolphins Panamanian fossil South Asian river dolphins Yangtze river dolphins	*Inia geoffrensis* *Isthminia panamensis* *Platanista gangetica* *Lipotes vexillifer*
	Sperm whales	Sperm whales Calvert Cliffs fossil	*Physeter macrocephalus* *Orycterocetus crocodilinus*

NOTES

PROLOGUE

1 long squeaks and moans on the record: The whalesong track on the interstellar golden records was mixed not with animal sounds, or world music, but with international greetings, something that says a lot about our relationship with these poorly known mammals. Timothy Ferris, "How the Voyager Golden Record Was Made," *New Yorker*, https://www.newyorker.com/tech/elements/voyager-golden-record-40th-anniversary-timothy-ferris.

2 perhaps understand their otherworldly: The premise, in fact, of the science fiction film *Star Trek IV: The Voyage Home*.

1. HOW TO KNOW A WHALE

10 By some measures: There are many different species of krill throughout the world's oceans, but the one in Wilhelmina Bay is Antarctic krill (*Euphausia superba*). It has been estimated that baleen whales in the Southern Ocean consume around two million metric tons of krill per austral summer season; it has been calculated that the major group of bony fish (Perciformes) in the Southern Ocean consume as much krill as all of the whales, penguins, and fur seals combined. The total amount of krill fluctuates by season and year, with the current best estimate of standing krill biomass in the Southern Ocean at 389 million metric tons. See Steve Reilly et al., "Biomass and Energy Transfer to Baleen Whales in the South Atlantic Sector of the Southern Ocean," *Deep Sea Research Part II: Topical Studies in Oceanography* 51 (2004): 1397–1409; Simeon Hill et al., "A Compilation of Parameters for Ecosystem Dynamics Models of the Scotia Sea-Antarctic Peninsula Region," *CCAMLR Science* 14 (2007): 1–25; and the Commission for the Conservation of Antarctic Marine Living Resources (CCAMLR) website at https://www.ccamlr.org/en/fisheries/krill-fisheries.

11 along the Antarctic Peninsula: Two of many examples: Ari S. Friedlaender et al., "Whale Distribution in Relation to Prey Abundance and Oceanographic Processes in Shelf Waters of the Western Antarctic

Peninsula," *Marine Ecology Progress Series* 317 (2006): 297–310; and Ari S. Friedlaender et al., "Extreme Diel Variation in the Feeding Behavior of Humpback Whales Along the Western Antarctic Peninsula During Autumn," *Marine Ecology Progress Series* 494 (2013): 281–89.

14 **organismal movement biologging:** For a valuable perspective on the history of the field, written by one of its founders, see Gerald L. Kooyman, "Genesis and Evolution of Bio-logging Devices: 1963–2002," *Memoirs of the National Institute of Polar Research* 58 (2004): 15–22. The field of biologging has progressed in remarkable ways since Kooyman's review; see, for example, Nicholas L. Payne et al., "From Physiology to Physics: Are We Recognizing the Flexibility of Biologging Tools?" *Journal of Experimental Biology* 217 (2014): 317–22.

15 **oftentimes in coordinated attacks:** See, for example, David Wiley et al., "Underwater Components of Humpback Whale Bubble-Net Feeding Behaviour," *Behaviour* 148 (2011): 575–602.

16 **most massive species of vertebrates:** Rorqual species are all closely related to one another, but it's unlikely that they form a family in the taxonomic sense, as explained in Michael R. McGowen et al., "Divergence Date Estimation and a Comprehensive Molecular Tree of Extant Cetaceans," *Molecular Phylogenetics and Evolution* 53 (2009): 891–906. Rorqual whales are, however, the most massive vertebrates ever. For sauropod dinosaur body sizes, see Mark Hallett and Mathew J. Wedel, *The Sauropod Dinosaurs: Life in the Age of Giants* (Baltimore: Johns Hopkins University Press, 2016). For rorqual body sizes, see Christina Lockyer, "Body Weights of Some Species of Large Whales," *Journal de Conseil International pour l'Exploration de la Mer* 36 (1976): 259–73. See also chapter 8 for a full discussion of the largest body sizes in blue whales.

17 **hemispheres over the seasons:** Data from the early-twentieth-century whaling industry supplied much of the basic evidence for large-scale migration patterns of baleen whales. See E. J. Slijper, *Whales* (London: Hutchinson, 1962), and David E. Gaskin, *The Ecology of Whales and Dolphins* (New York: Heinemann, 1982).

17 **still retain olfactory lobes:** Stephen J. Godfrey et al., "On the Olfactory Anatomy in an Archaic Whale (Protocetidae, Cetacea) and the Minke Whale *Balaenoptera acutorostrata* (Balaenopteridae, Cetacea)," *Anatomical Record* 296 (2013): 257–72.

17 **amazing process of a lunge:** Jeremy A. Goldbogen, "The Ultimate Mouthful: Lunge Feeding in Rorqual Whales," *American Scientist* 98 (2010): 124–31.

18 events on the planet: The first publication to describe lunge feeding as the largest biomechanical event on Earth was Paul F. Brodie, "Noise Generated by the Jaw Actions of Feeding Fin Whales," *Canadian Journal of Zoology* 71 (1993): 2546–50. Paul was instrumental in conducting fieldwork in Iceland (see chapter 10), assisting with anatomical research both in the 1980s and when Bob returned to Iceland with his own lab, along with Jeremy and me. For thorough data on engulfment volumes, see Jeremy A. Goldbogen et al., "Scaling of Lunge Feeding Performance in Rorqual Whales: Mass-Specific Energy Expenditure Increases with Body Size and Progressively Limits Diving Capacity," *Functional Ecology* 26 (2012): 216–26.

18 in mirrored unison: Colin Ware et al., "Shallow and Deep Lunge Feeding of Humpback Whales in Fjords of the West Antarctic Peninsula," *Marine Mammal Science* 27 (2011): 587–605.

18 roll their bodies to feed: Ari S. Friedlaender et al., "Three-Dimensional Context Dependence of Divergent Lateralized Feeding Strategies in Blue Whales," *Current Biology* 27 (2017): R1206–8.

19 with a patch of krill: The footage in the paper is well worth viewing, but the paper itself is short and sweet: Jeremy A. Goldbogen et al., "Underwater Acrobatics by the World's Largest Predator: 360° Rolling Maneuvers by Lunge Feeding Blue Whales," *Biology Letters* 9 (2013): 20120986.

19 Peninsula to subtropical waters: Jessica F. Lee et al., "Behavior of Satellite-Tracked Antarctic Minke Whales (*Balaenoptera bonaerensis*) in Relation to Environmental Factors Around the Western Antarctic Peninsula," *Animal Biotelemetry* 5 (2017): 23.

19 records for a mammal: Gregory S. Schorr et al., "First Long-Term Behavioral Records from Cuvier's Beaked Whales (*Ziphius cavirostris*) Reveal Record-Breaking Dives," *PLoS ONE* 9 (2014): e92633.

19 taken from biopsy darts: Biopsy darts used on whales consist of hollow steel bolts shot from crossbows or compressed-air guns that collect pinkie finger–size slugs of tissue. The skin and fatty tissue in these samples can reveal a tremendous amount about the diet, reproductive history, and genetics of an individual whale. For an important early paper on the use of biopsy darts on whales, see Richard H. Lambertsen, "A Biopsy System for Large Whales and Its Use for Cytogenetics," *Journal of Mammalogy* 68 (1987): 443–45.

19 austral summer to feed: Kristin Rasmussen et al., "Southern Hemisphere Humpback Whales Wintering off Central America: Insights from Water Temperature into the Longest Mammalian Migration," *Biology Letters* 3 (2007): 302.

21 **part of the community:** Perhaps the only guide to past whale communities in the Antarctic Peninsula is an account from just after the peak of whaling operations in the area: Stanley Kemp and A. G. Bennett, "On the Distribution and Movements of Whales on the South Georgia and South Shetland Whaling Grounds," *Discovery Reports* 6 (1932): 165–90.

22 **Southern Hemisphere alone:** According to the most recent and most comprehensive tally, over two million of the more than three million whales killed in the twentieth century were from the Southern Ocean. Robert C. Rocha Jr. et al., "Emptying the Oceans: A Summary of Industrial Whaling Catches in the 20th Century," *Marine Fisheries Review* 76 (2014): 37–48.

22 **baleen whales ever recorded:** The number of humpback whales in Wilhelmina Bay was extrapolated from sightings along a transect. See Douglas P. Nowacek et al., "Super-aggregations of Krill and Humpback Whales in Wilhelmina Bay, Antarctic Peninsula," *PLoS ONE* 6 (2011): e19173.

22 **start of the twentieth century:** See Alyson H. Fleming and Jennifer Jackson, "Global Review of Humpback Whales (*Megaptera novaeangliae*)," *NOAA Technical Memorandum* NOAA-TM-NMFS-SWFSC-474 (2011): 1–206.

24 **first scientists to visit:** Adrien de Gerlache discovered and named Wilhelmina Bay during the Belgian Antarctic expedition in 1898. Gerlache's crew aboard the *Belgica* included several scientists, which was not necessarily typical of expeditions to Antarctica prior to the twentieth century. Part of what makes the history of Antarctic exploration so fascinating is the primacy of geopolitical motives over scientific justifications, especially in the competition to claim unoccupied territory. See David Day, *Antarctica: A Biography* (Oxford, UK: Oxford University Press, 2013).

2. MAMMALS LIKE NO OTHER

25 **the range of blue whales:** An authoritative and handy book for such maps is Brent Stewart et al., *National Audubon Society Guide to Marine Mammals of the World* (New York: Knopf, 2002). Casual browsing will reveal so much that we don't know about the world's whales.

26 **no threat to our lives:** Humans are not whale prey; whales are fundamentally uninterested in eating us, although humans have been killed by captive killer whales in aquariums and certainly during whaling hunts,

especially prior to the late nineteenth century. See Carl Safina, *Beyond Words: What Animals Think and Feel* (New York: Henry Holt, 2015).

27 **that reaches into Deep Time:** Deep Time is an idea born out of the geological sciences with strong roots in the early nineteenth century. Martin J. S. Rudwick has written extensively on many episodes of the history of this idea; a particularly useful starting point may be Martin J. S. Rudwick, *Earth's Deep History: How It Was Discovered and Why It Matters* (Chicago: University of Chicago Press, 2014).

27 **some features of these past worlds:** See James Zachos et al., "Trends, Rhythms, and Aberrations in Global Climate 65 Ma to Present," *Science* 292 (2001): 686–93; and Bärbel Hönisch et al., "The Geological Record of Ocean Acidification," *Science* 335 (2012): 1058–63.

27 **obviously, visibly mammalian:** Much of the evidence for the early evolution of whales is documented in the fossil record. The species descriptions are pulled from a body of primary literature that can be tracked in the following summaries: Philip D. Gingerich, "Evolution of Whales from Land to Sea: Fossils and a Synthesis," in *Great Transformations: Essays in Honor of Farish A. Jenkins*, ed. K. P. Dial, N. H. Shubin, and E. Brainerd (Chicago: University of Chicago Press, 2015), pp. 239–56; and J. G. M. Thewissen, *The Walking Whales: From Land to Water in Eight Million Years* (Berkeley: University of California Press, 2014).

28 **area that is now Pakistan:** Philip D. Gingerich et al., "Origin of Whales in Epicontinental Remnant Seas: New Evidence from the Early Eocene of Pakistan," *Science* 220 (1983): 403–6.

28 **"ambulatory, or walking, whale":** J. G. M. Thewissen et al., "Fossil Evidence for the Origin of Aquatic Locomotion in Archaeocete Whales," *Science* 263 (1994): 210–11.

28 **the mother whale:** Philip D. Gingerich et al., "New Protocetid Whale from the Middle Eocene of Pakistan: Birth on Land, Precocial Development, and Sexual Dimorphism," *PLoS ONE* 4 (2009): e4366.

29 **rolled like a tiny conch shell:** For more details about ear anatomy in whales, see Zhe-Xi Luo and Philip D. Gingerich, "Terrestrial Mesonychia to Aquatic Cetacea: Transformation of the Basicranium and Evolution of Hearing in Whales," *University of Michigan Papers on Paleontology* 31 (1999): 1–98; and J. G. M. Thewissen and Sirpa Nummela, *Sensory Evolution on the Threshold: Adaptations in Secondarily Aquatic Vertebrates* (Berkeley: University of California Press, 2008).

30 **million years, until today:** The oldest fossil whale species is nearly 53 million years old, but 50 million years is a reasonably close figure for the purposes of a general discussion; see more precision in Nicholas D. Pyenson, "The Ecological Rise of Whales Chronicled by the Fossil Record," *Current Biology* 27 (2017): R558–64.

31 **new dimension of adaptation:** George Gaylord Simpson, a vertebrate paleontologist and one of the chief architects of the modern synthesis of biology in the midtwentieth century, explicitly stated this idea as an adaptive zone: penguins, for example, evolved into a new adaptive zone when they achieved underwater, forelimb-propelled swimming, which was unlike any other avian lineage before it. It's a credit to the durability of an idea that biologists still use terminology developed decades ago. See George G. Simpson, *Tempo and Mode in Evolution* (New York: Columbia University Press, 1944); and Dolph Schluter, *The Ecology of Adaptive Radiation* (Oxford, UK: Oxford University Press, 2000).

31 **to living in oceanic ones:** Marine mammals and marine reptiles represent separate branches on the family tree of four-limbed vertebrates (called tetrapods). For a comprehensive overview of these multiple invasions of the oceans, see Neil P. Kelley and Nicholas D. Pyenson, "Evolutionary Innovation and Ecology in Marine Tetrapods from the Triassic to the Anthropocene," *Science* 301 (2015): aaa3716.

32 **any other mammal:** Manatees, dugongs, and sea cows belong to a group of herbivorous marine mammals known as sirenians. This group descends from land ancestors with four weight-bearing limbs, which lived in the Caribbean about fifty million years ago. Early sirenian evolution shows many parallels with stages of early cetacean evolution, including the loss of hind limbs. See Daryl P. Domning, "Sirenian Evolution," in Bernd Würsig et al., eds., *Encyclopedia of Marine Mammals*, 3rd ed. (San Diego, CA: Academic Press/Elsevier, 2018), pp. 856–59. For a broad overview of the many returns of marine mammals to the sea, see Annalisa Berta, *The Rise of Marine Mammals: 50 Million Years of Evolution* (Baltimore: Johns Hopkins University Press, 2017).

32 **from the neck down:** Philip D. Gingerich, "Evolution of Whales from Land to Sea," *Proceedings of the American Philosophical Society* 156 (2012): 309–23.

32 **connected in life by webbing:** For a description of the discovery of *Ambulocetus*, see J. G. M. Thewissen, *The Walking Whales: From Land to Water in Eight Million Years* (Berkeley: University of California Press, 2014).

33 **using lift, instead of drag:** A seminal paper that remains up-to-date regarding the transitional states of the possible locomotory modes for early whales is J. G. M. Thewissen and F. E. Fish, "Locomotor Evolution in the Earliest Cetaceans: Functional Model, Modern Analogues, and Paleontological Evidence," *Paleobiology* 23 (1997): 482–90.

33 **tails might have moved:** Gingerich, 2012.

33 **possible evolutionary relationship:** William Turner, along with William Henry Flower (then at the British Museum at South Kensington) made these observations at the end of the nineteenth century, although Turner's publication came many years later. William Turner, *The Marine Mammals in the Anatomical Museum of the University of Edinburgh* (London: Macmillan, 1912).

34 **deepest origins of whales:** Christian de Muizon, "Walking with Whales," *Nature* 413 (2001): 259–60.

34 **groups of paleontologists:** These two papers were published the same week in 2001: J. G. M. Thewissen et al., "Skeletons of Terrestrial Cetaceans and the Relationship of Whales to Artiodactyls," *Nature* 413 (2001): 277–81; and Philip D. Gingerich et al., "Origin of Whales from Early Artiodactyls: Hands and Feet of Eocene Protocetidae from Pakistan," *Science* 293 (2001): 2239–42.

35 **perhaps for eating clams:** Philip D. Gingerich et al., "*Makaracetus bidens*, a New Protocetid Archaeocete (Mammalia, Cetacea) from the Early Middle Eocene of Balochistan (Pakistan)," *University of Michigan Publications in Paleontology* 31 (2005): 197–210.

35 **like a lost dog:** At the 2006 meeting of the Society of Vertebrate Paleontology in Ottawa, Canada, I remember Philip Gingerich giving a talk in which he used a similar phrasing to urge a bit more honesty in paleontological artistic reconstructions—that it's a mistake to depict *Pakicetus* as something like a little lost dog when we really possess so little of its skeleton to make an informed reconstruction of what it looked like.

36 **in one lineup:** See a wonderful reconstruction by Carl Buell, figure 2 in J. G. M. Thewissen and Ellen M. Williams, "The Early Radiations of Cetacea (Mammalia): Evolutionary Pattern and Developmental Correlations," *Annual Review of Ecology and Systematics* 33 (2002): 73–90.

37 **skull of any vertebrate animal:** Skull anatomy and osteology is a rite of passage for students of vertebrate morphology. For a deep dive, see James Hanken and Brian K. Hall, eds., *The Skull: Functional and*

Evolutionary Mechanisms, Vol. 3 (Chicago: University of Chicago Press, 1993).

37 **skull of a bottlenose dolphin:** James G. Mead and R. Ewan Fordyce, "The Therian Skull: A Lexicon with Emphasis on the Odontocetes," *Smithsonian Contributions to Zoology* 627 (2009): 1–248.

38–39 **nostrils positioned more forward:** Domning, 2017.

39 **strapped to the dolphin's forehead:** Whitlow W. L. Au, *The Sonar of Dolphins* (New York: Springer Science & Business Media, 2012).

40 **that do it underwater:** David R. Lindberg and Nicholas D. Pyenson, "Things That Go Bump in the Night: Evolutionary Interactions Between Cephalopods and Cetaceans in the Tertiary," *Lethaia* 40 (2007): 335–43.

40 **pinpoint the source:** Kenneth S. Norris and George W. Harvey, "Sound Transmission in the Porpoise Head," *Journal of the Acoustical Society of America* 56 (1974): 659–64; Ted Cranford et al., "Functional Morphology and Homology in the Odontocete Nasal Complex: Implications for Sound Generation," *Journal of Morphology* 228 (1996): 223–85.

40 **the time of *Ambulocetus*:** Sirpa Nummela et al., "Eocene Evolution of Whale Hearing," *Nature* 430 (2004): 776–78.

40 **are not easily procured:** W. W. Au, "History of Dolphin Biosonar Research," *Acoustics Today* (Fall 2015): 10–17.

41 **find a small object:** Kenneth S. Norris et al., "An Experimental Demonstration of Echo-Location Behavior in the Porpoise, *Tursiops truncatus* (Montagu)," *Biological Bulletin* 120 (1961): 163–76.

41 **and even ear bones:** Megan F. McKenna et al., "Morphology of the Odontocete Melon and Its Implications for Acoustic Function," *Marine Mammal Science* 28 (2012): 690–713.

41 **questions in biology today:** Neil Shubin et al., "Deep Homology and the Origins of Evolutionary Novelty," *Nature* 457 (2009): 818–23.

41 **their dinosaur ancestors:** Richard O. Prum and Andrew H. Brush, "Which Came First, the Feather or the Bird?" *Scientific American* 288 (2003): 84–93.

41 **shoulders tucked firmly inside:** Tyler R. Lyson et al., "Evolutionary Origin of the Turtle Shell," *Current Biology* 23 (2013): 1113–19.

42 **none of their forebears:** Pyenson, 2017.

3. THE STORIES BONES TELL

47 **A shark bite:** Shark-bitten whale bones are not uncommon at fossil-rich sites around the world. See Thomas A. Deméré and Richard A. Cerutti,

"A Pliocene Shark Attack on a Cethotheriid Whale," *Journal of Paleontology* 56 (1982): 1480–82; and Dana J. Ehret et al., "Caught in the Act: Trophic Interactions Between a 4-Million-Year-Old White Shark (*Carcharodon*) and Mysticete Whale from Peru," *Palaios* 24 (2009): 329–33.

49 **a gap in the fossil record:** See Gingerich, 2012.

50 **a single layer of rock:** Raymond R. Rogers et al., eds., *Bonebeds: Genesis, Analysis, and Paleobiological Significance* (Chicago: University of Chicago Press, 2010).

50 **individual ear bones:** Remington Kellogg, "Pelagic Mammals from the Temblor Formation of the Kern River Region, California," *Proceedings of the California Academy of Sciences* 4 (1931): 217–397.

51 **bones for any mammal:** John D. Currey, "Mechanical Properties of Bone Tissues with Greatly Differing Functions," *Journal of Biomechanics* 12 (1979): 313–19.

51 **herbivores called desmostylians:** Mark T. Clementz et al., "A Paleoecological Paradox: The Habitat and Dietary Preferences of the Extinct Tethythere *Desmostylus*, Inferred from Stable Isotope Analysis," *Paleobiology* 29 (2003): 506–19.

51 **more interested in its fossil sea turtles:** See Shannon C. Lynch and James F. Parham, "The First Report of Hard-shelled Sea Turtles (Cheloniidae *sensu lato*) from the Miocene of California, Including a New Species (*Euclastes hutchisoni*) with Unusually Plesiomorphic Characters," *PaleoBios* 23 (2003): 21–35.

52 **condensed interval of time:** The Sharktooth Hill bonebed has long been known to produce bits of fossil whales (and other extinct creatures) in the foothills of the Central Valley, near Bakersfield. Paleontologists have collected many thousands of shark teeth—whose hard, slate-blue enamel and rich diversity literally erode from the hillsides in plain sight, giving the bonebed its name—there since the nineteenth century. By the twentieth century, oil exploration brought intense scientific focus on the age and contents of the underlying petroleum-laden rocks around Bakersfield, and with it sufficient interest in the identity of the fossils nearby. The University of California Museum of Paleontology, as well as the Natural History Museum of Los Angeles County, have drawers of fossils from the bonebed still awaiting careful inspection. For a summary of findings about the geology of the Sharktooth Hill bonebed, see Nicholas D. Pyenson et al., "Origin of a Widespread Marine Bonebed Deposited During the Middle Miocene Climatic Optimum," *Geology* 37 (2009): 519–22.

52 **attention to fossil whales:** Remington Kellogg, "A Review of the Archaeoceti," *Carnegie Institution of Washington Publication* 482 (1936): 1–366.

53 **came from its tail:** Philip D. Gingerich et al., "Hind Limbs of Eocene *Basilosaurus*: Evidence of Feet in Whales," *Science* 249 (1990): 154–57.

53 **about six tons:** Philip D. Gingerich, "Body Weight and Relative Brain Size (Encephalization) in Eocene Archaeoceti (Cetacea)," *Journal of Mammalian Evolution* 23 (2016): 17–31.

54 **earliest elephant relatives:** Philip D. Gingerich, "Marine Mammals (Cetacea and Sirenia) from the Eocene of Gebel Mokattam and Fayum, Egypt: Stratigraphy, Age and Paleoenvironments," *University of Michigan Museum of Paleontology, Papers on Paleontology* 30 (1992): 1–84.

54–55 **mammal, living or extinct, including hyenas:** Eric J. Snively, "Bone-Breaking Bite Force of *Basilosaurus isis* (Mammalia, Cetacea) from the Late Eocene of Egypt Estimated by Finite Element Analysis," *PLoS ONE* 10 (2015): e0118380.

55 **killer whales do today:** Mark D. Uhen, "Form, Function, and Anatomy of *Dorudon atrox* (Mammalia, Cetacea): An Archaeocete from the Middle to Late Eocene of Egypt," *University of Michigan Museum of Paleontology, Papers on Paleontology* 34 (2004): 1–222.

55 **reflect ancient shorelines:** Shanan E. Peters et al., "Sequence Stratigraphic Control on Preservation of Late Eocene Whales and Other Vertebrates at Wadi Al-Hitan, Egypt," *Palaios* 24 (2009): 290–302.

4. TIME TRAVEL ON THE FOSSIL WHALE HIGHWAY

59 **Chile, near Concepción:** There is a vast literature to consult about Darwin's travels on the HMS *Beagle*. Darwin's account of his time aboard the ship, under the command of Captain Robert FitzRoy, and its expedition around the world has been published in several iterations and in separate books. Today these works are all loosely grouped under the title *Voyage of the Beagle*, which represents a condensation of Darwin's narrative corpus on the expedition, mostly edited after his death. For Darwin's account of his time in Chiloé and Concepción, I consulted chapters 15 and 16 in his original 1839 publication on that part of the expedition: Charles R. Darwin, *Narrative of the Surveying Voyages of His Majesty's Ships Adventure and Beagle Between the Years 1826 and 1836, Describing Their Examination of the Southern Shores of South America, and the Beagle's Circumnavigation of the Globe: Journal and Remarks, 1832–1836* (London: Henry Colburn,

1839). The original text is conveniently available online through open access from Darwin Online, supported by Cambridge University; see John van Wyhe, ed., *The Complete Work of Charles Darwin Online* (2002), http://darwin-online.org.uk/.

59 **idea of plate tectonics:** David Whitehouse, *Into the Heart of Our World: A Journey to the Center of the Earth: A Remarkable Voyage of Scientific Discovery* (New York: Pegasus Books, 2016).

59 **over geologic time:** For a history of plate tectonics, see Naomi Oreskes, *Plate Tectonics: An Insider's History of the Modern Theory of the Earth* (Boulder, CO: Westview Press, 2003); and William Glen, *Road to Jaramillo: Critical Years of the Revolution in Earth Science* (Redwood City, CA: Stanford University Press, 1982).

60 **the South American cone:** See Rudwick, 2014.

60 **land mammals in Patagonia:** For a summary of Darwin's geologizing in South America, the first volume of Janet Browne's two-volume set provides a definitive synthesis: E. Janet Browne, *Charles Darwin, Vol. 1: Voyaging* (London: Jonathan Cape, 1995).

60 **near the town of Caldera:** Darwin recorded his itinerary and geologizing in the Atacama in chapter 18 of his *Narrative of the Surveying Voyages.*

60–61 **including a national museum:** For more on Darwin's social network in South America, especially with respect to its impacts on Chilean institutions, see Patience A. Schell, *The Sociable Sciences: Darwin and His Contemporaries in Chile* (New York: Palgrave Macmillan, 2013).

61 **the Humboldt Current:** G. Hempel and K. Sherman, *Large Marine Ecosystems of the World* (Amsterdam: Elsevier, 2003).

62 **penguins all want to be:** For a seminal description of the impact of upwelling on cetaceans, see Donald A. Croll et al., "From Wind to Whales: Trophic Links in a Coastal Upwelling System," *Marine Ecology Progress Series* 289 (2005): 117–30.

62 **beneath the Earth's surface:** Jan Zalasiewicz, *The Planet in a Pebble: A Journey into Earth's Deep History* (Oxford, UK: Oxford Landmark Science, 2012).

63 **never complete skeletons:** See R. A. Philippi, *Los fósiles terciarios i cuartarios de Chile* (Leipzig: F. A. Brockhaus, 1887).

63 **school bus–size predatory sharks:** For fossil seabirds, see Gerald Mayr and David Rubilar-Rogers, "Osteology of a New Giant Bony-Toothed Bird from the Miocene of Chile, with a Revision of the Taxonomy of Neogene Pelagornithidae," *Journal of Vertebrate Paleontology* 30 (2010): 1313–30. For aquatic sloths, see Eli Amson et al.,

"Gradual Adaptation of Bone Structure to Aquatic Lifestyle in Ex-
tinct Sloths from Peru," *Proceedings of the Royal Society of London
B: Biological Sciences* 281 (2014): 20140192. For megatoothed sharks,
see Catalina Pimiento et al., "Geographical Distribution Patterns of
Carcharocles megalodon over Time Reveal Clues About Extinction
Mechanisms," *Journal of Biogeography* 43 (2016): 1645–55.

63 **fall of each species:** Much of this work formed the basis of an honors
thesis at Universidad de Chile by Catalina Carreño. Parts of her findings on the stratigraphy of the Caldera Basin have since been published in J. P. Le Roux et al., "Oroclinal Bending of the Juan Fernández
Ridge Suggested by Geohistory Analysis of the Bahía Inglesa Formation, North-Central Chile," *Sedimentary Geology* 333 (2016): 32–49.

64 **true for some of them:** Robert M. Norris and Robert W. Webb, *Geol-
ogy of California*, 2nd ed. (Hoboken, NJ: John Wiley and Sons, 1990).
Robert Norris's brother was, by chance, Kenneth Norris, one of the
more influential marine mammalogists of the twentieth century. See
Randall Jarrell, ed., *Kenneth S. Norris: Naturalist, Cetologist and
Conservationist, 1924–1998: An Oral History Biography* (Berkeley:
University of California Press in association with University Library,
UC Santa Cruz, 2010), https://escholarship.org/uc/item/5kf1t3wg.

65 **back in a laboratory:** For background on the use of zircon grains in
radiometric dating (especially for uranium-lead dating), see Zalasiewicz, 2012.

68 **part of the Atacama:** Cerro Ballena is located not more than a few
hours' drive away from the San José copper-gold mine in the Atacama Desert, which suffered a cave-in on August 5, 2010, trapping
thirty-three miners more than two thousand feet underground for
sixty-nine days. All thirty-three miners were rescued. This mining
accident was the focus of an excellent *NOVA* PBS television broad-
cast, "Emergency Mine Rescue" (2010).

5. THE AFTERLIFE OF A WHALE

72 **a pathway of decay:** See figure 3 in Anna K. Behrensmeyer and Susan
M. Kidwell, "Taphonomy's Contributions to Paleobiology," *Paleobiology* 11 (1985): 105–19.

72 **Aristotle to casual spectators:** Armand Marie Leroi, *The Lagoon: How
Aristotle Invented Science* (New York: Bloomsbury Publishing, 2014).

72 **exploding whale videos on YouTube:** In November 1970 near Florence,
Oregon, the Oregon Highway Division tried to remove a dead sperm

whale from the shoreline using dynamite. While they intended for sea-birds to eat the blasted remains, the result was more than merely aero-solized whale gristle. The detonation rocketed hundred-pound chunks of whale carcass up to eight hundred feet away, smashing parked cars and endangering all inside the blast zone—all in front of television cameras. See https://www.youtube.com/watch?v=uD5sPgV61bw.

73 knowledge about whales: John Hunter and Joseph Banks, "Observations on the Structure and Oeconomy of Whales by John Hunter, Esq. F. R. S.; Communicated by Sir Joseph Banks, Bart. P. R. S.," *Philosophical Transactions of the Royal Society of London* 77 (1787): 371–450. See also William Scoresby Jr., "Account of the *Balaena mysticetus*, or Great Northern Greenland Whale," *Memoirs of the Wernerian Natural History Society* 1 (1810): 578–86; and William H. Flower, "On a Lesser Fin-Whale (*Balaenoptera rostrata*, Fabr.) Recently Stranded on the Norfolk Coast," *Proceedings of the Zoological Society of London* (1864): 252–58.

74 cow or a tax collector: Turner, 1912. Also see Richard Owen, *On the Anatomy of Vertebrates*, vols. 1–3 (London: Longman, Green, 1866–1868).

74 description as a new species: William Turner, "An Account of the Great Finner Whale (*Balaenoptera sibbaldii*) Stranded at Longniddry," *Transactions of the Royal Society of Edinburgh* 26 (1870): 197–251.

75 for Frederick William True: Frederick W. True, *The Whalebone Whales of the Western North Atlantic Compared with Those Occurring in European Waters*, Smithsonian Contributions to Knowledge 33 (1904): 1–332.

75 some species of beaked whales: Kirsten Thompson et al., "The World's Rarest Whale," *Current Biology* 22 (2012): R905–6.

76 deploy properly in life: Robert E. Shadwick et al., "Novel Muscle and Connective Tissue Design Enables Hyper-extensibility and Controls Engulfment Volume in Lunge-Feeding Rorqual Whales," *Journal of Experimental Biology* 216 (2013): 2691–701.

76 off the California coast: Tom Vetter, *30,000 Leagues Undersea: True Tales of a Submariner and Deep Submergence Pilot* (Tom Vetter Books, 2015), www.tomvetterbooks.com.

76 called them whalefalls: Craig R. Smith and Amy R. Baco "Ecology of Whale Falls at the Deep-Sea Floor," *Oceanography and Marine Biology: An Annual Review* 41 (2003): 311–54; and Craig R. Smith et al., "Whale-fall Ecosystems: Recent Insights into Ecology, Paleoecology, and Evolution," *Annual Review of Marine Science* 7 (2015): 571–96.

78 depends on whalefall skeletons: Greg W. Rouse et al., "*Osedax*: Bone-Eating Marine Worms with Dwarf Males," *Science* 305 (2004): 668–71.

79 itself by dissolving it: Also, as a side note, you'll never see a male *Osedax* because they don't develop past the larval stage, and they're housed by the dozens inside the body of the female, which is rooted into the bone.

79 to leap to another: The distinctive burrows that *Osedax* leaves inside whale bone have since been discovered in fossil whales, as well as in fossil bird, fossil sea turtle, and even plesiosaur bones. That breadth of host species shows that *Osedax* has been successful at colonizing a vast range of fallen skeletons, not just mammalian ones, on the seafloor for over 100 million years—well before the first whales. Silvia Danise and Nicholas D. Higgs, "Bone-Eating *Osedax* Worms Lived on Mesozoic Marine Reptile Deadfalls," *Biology Letters* 11 (2015): 20150072.

79 to appear on the seafloor: Craig R. Smith et al., "Whale-fall Ecosystems: Recent Insights into Ecology, Paleoecology, and Evolution," *Annual Review of Marine Science* 7 (2015): 571–96.

80 to a fossil whalefall: Nicholas D. Pyenson and David M. Haasl, "Miocene Whale-fall from California Demonstrates That Cetacean Size Did Not Determine the Evolution of Modern Whale-fall Communities," *Biology Letters* 3 (2007): 709–11.

81 half of the twentieth century: Ronald Rainger, "Everett C. Olson and the Development of Vertebrate Paleoecology and Taphonomy," *Archives of Natural History* 24 (1997): 373–96.

81 shores of the North Sea: Wilhelm Schäfer, *Ecology and Palaeoecology of Marine Environments* (Chicago: University of Chicago Press, 1972).

81 from the rest of the carcass: Johannes Weigelt and Judith Schaefer, *Recent Vertebrate Carcasses and Their Paleobiological Implications* (Chicago: University of Chicago Press, 1989).

82 coordinated by government agencies: Prior to 1972, there were no federal laws in the United States to guide the collection of marine mammal stranding data along American coastlines. Although the Marine Mammal Protection Act provided the legal rubric for creating stranding networks, it wasn't until 1991 that network reporting achieved decent coverage across U.S. shorelines. Data from all regions of the marine mammal stranding network are published annually for the National Oceanic and Atmospheric Administration (NOAA) in the Marine Mammal Protection Act Annual Report Archive.

82 whale species from boats: Jay Barlow and Karin A. Forney, "Abundance and Population Density of Cetaceans in the California Current Ecosystem," *Fishery Bulletin* 105 (2007): 509–26.

82 The answer: surprisingly well: Nicholas D. Pyenson, "Carcasses on the Coastline: Measuring the Ecological Fidelity of the Cetacean Stranding Record in the Eastern North Pacific Ocean," *Paleobiology* 36 (2010): 453–80.

83 piles of fossil ambergris: Angela Baldanza et al., "Enigmatic, Biogenically Induced Structures in Pleistocene Marine Deposits: A First Record of Fossil Ambergris," *Geology* 41 (2013): 1075–78.

83 found together in a sand berm: Catherine Kemper et al., "Subfossil Evidence of Strandings of the Sperm Whale *Physeter macrocephalus* in Gulf St. Vincent, South Australia," *Records of the South Australia Museum* 29 (1997): 41–53.

6. ROCK PICKS AND LASERS

93 impending northbound lane: Abigail Tucker, "Save the Whalebones," *Smithsonian*, June 2012, 84–85.

95 or a 3-D printer: Carl Zimmer, "Laser Cowboys and the Fossils of the Future," *Popular Mechanics*, May 2014, 64–69. Also see Michael Weinberg, "It Will Be Awesome If They Don't Screw It Up: 3D Printing, Intellectual Property, and the Fight over the Next Great Disruptive Technology" (white paper, Public Knowledge, November 2010); and Michael Weinberg, "What's the Deal with Copyright and 3D Printing?" (white paper, Public Knowledge Institute for Emerging Innovation, January 2013).

7. CRACKING THE CASE OF CERRO BALLENA

99 What happened at Cerro Ballena: Nicholas D. Pyenson et al., "Repeated Mass Strandings of Miocene Marine Mammals from Atacama Region of Chile Point to Sudden Death at Sea," *Proceedings of the Royal Society B: Biological Sciences* 281 (2014): 20133316.

103 consumed other marine mammals: Olivier Lambert et al., "The Giant Bite of a New Raptorial Sperm Whale from the Miocene Epoch of Peru," *Nature* 466 (2010): 105–8.

103 to the north, in Peru: Christian de Muizon, "Walrus-Like Feeding Adaptation in a New Cetacean from the Pliocene of Peru," *Nature* 365 (1993): 745–48.

103 **found only in South America:** Christian de Muizon and H. Gregory McDonald, "An Aquatic Sloth from the Pliocene of Peru," *Nature* 375 (1995): 224–27; and Christian de Muizon et al., "The Evolution of Feeding Adaptations of the Aquatic Sloth *Thalassocnus*," *Journal of Vertebrate Paleontology* 24 (2004): 398–410.

103 **seventeen-foot wingspan:** Daniel T. Ksepka, "Flight Performance of the Largest Volant Bird," *Proceedings of the National Academy of Sciences* 111 (2014): 10624–29.

104 **100 million to 70 million years ago:** Simon N. Jarman, "The Evolutionary History of Krill Inferred from Nuclear Large Subunit rDNA Sequence Analysis," *Biological Journal of the Linnean Society* 73 (2001): 199–212.

105 **around three million years ago:** Catalina Pimiento and Christopher F. Clements, "When Did *Carcharocles megalodon* Become Extinct? A New Analysis of the Fossil Record," *PLoS ONE* 9 (2014): e111086.

106 **stranding had a putative explanation:** See Pyenson et al., 2014; and Aleta Hohn et al., "Report on Marine Mammal Unusual Mortality Event UMESE0501Sp: Multispecies Mass Stranding of Pilot Whales (*Globicephala macrorhynchus*), Minke Whale (*Balaenoptera acutorostrata*), and Dwarf Sperm Whales (*Kogia sima*) in North Carolina on 15–16 January 2005, *NOAA Technical Memorandum* NMFS-SEFSC-537 (2006): 1–222.

107 **duration of a bloom:** The Andes are among the most iron-rich mountain ranges in the world, providing abundant source rock for iron runoff to the Pacific Ocean. There is evidence that iron runoff has boosted coastal productivity in waters directly off the Atacama, less than one hundred miles away from Cerro Ballena; see Laurent Dezileau et al., "Iron Control of Past Productivity in the Coastal Upwelling System off the Atacama Desert, Chile," *Paleoceanography* 19 (2004): PA3012. The coincidences of iron runoff with increases in coastal productivity in an upwelling system are exactly the right kinds of conditions to promote harmful algal blooms.

108 **awful way that algal poisons:** See Donald M. Anderson, "Red Tides," *Scientific American* 271 (1994): 62–68; and Frances M. Van Dolah et al., "Impacts of Algal Toxins on Marine Mammals," in *Toxicology of Marine Mammals* (London: Taylor & Francis, 2003), pp. 247–69.

108 **billfishes were not immune:** Among many examples, see Leanne J. Flewelling et al., "Brevetoxicosis: Red Tides and Marine Mammal Mortalities," *Nature* 435 (2005): 755–56.

108 **on their backs, belly up:** One of the most compelling modern examples of the effects of harmful algal blooms on whales—including the specific behavioral, physiological, and taphonomic signs, happened over five weeks in Cape Cod Bay, Massachusetts, where more than a dozen humpback whales (and other rorquals) died from red tide poisoning. Some of the whales were observed on a whale watch hours before their death, brought about by the neurotoxicity of the red tides (the whales were poisoned by the mackerel that they ate, which were loaded with dinoflagellate toxins). A large proportion of the dead whales were stranded belly up. See Anderson, 1994.

109 **shells called a coquina:** This rock unit has revealed interesting patterns in the evolution of seal and sea lion communities in South America in the past few million years, which were described in Ana M. Valenzuela-Toro et al., "Pinniped Turnover in the South Pacific Ocean: New Evidence from the Plio-Pleistocene of the Atacama Desert, Chile," *Journal of Vertebrate Paleontology* 33 (2013): 216–23.

110 **probably not unique:** One of the key predictions generated from our team's work at Cerro Ballena was that baleen whale mass stranding events ought to become more frequent in the coming years as populations of these whale species recover from the devastation of industrial whaling in the twentieth century. A year after the publication of our Cerro Ballena paper, members of our team (including Caro) were recruited to investigate exactly this type of mass stranding, discovered in Patagonian Chile: a mass stranding of at least 343 baleen whales (likely all consisting of sei whales), stranded in a fjord system along the central part of Patagonia. This mass stranding event was the largest baleen whale mass stranding ever reported and very likely linked to harmful algal bloom poisoning. See Häussermann et al., "Largest Baleen Whale Mass Mortality During Strong El Niño Event Is Likely Related to Harmful Toxic Algal Bloom," *PeerJ* 5 (2017): e3123.

111 **what should be true:** Margaretha Brongersma-Sanders first articulated the importance of mass-mortality events at sea as semaphores of broader oceanographic processes in a classic paper with continuing relevance: Margaretha Brongersma-Sanders, "Mass Mortality in the Sea," *Geological Society of America Memoirs* 67 (1957): 941–1010. For the deep history perspective on harmful algal blooms over geologic time, see James W. Castle and John H. Rodgers Jr., "Hypothesis for the Role of Toxin-Producing Algae in Phanerozoic Mass Extinctions Based on Evidence from the Geologic Record and Modern Environments," *Environmental Geosciences* 16 (2009): 1–23.

111 with an Internet connection: Open access 3-D digital models of fossil whale skeletons from Cerro Ballena can be measured, manipulated, and downloaded from the Smithsonian's Digitization Program Office 3-D lab's website at https://3d.si.edu/.

111 than the first dinosaurs: David Rubilar Rogers et al., eds., *Dinosaurios de Chile, Pasado y Presente* (Santiago: Editorial SurCiencia Chile, 2017).

112 turn of the twentieth century: The Smithsonian Institution Archives maintains an excellent set of images of the Solar Observation Station at Mount Montezuma, which was located near the city of Calama, Chile, from the 1920s to the 1930s. For more about the history of astronomy in this part of South America, see Dava Sobel, *The Glass Universe: How the Ladies of the Harvard Observatory Took the Measure of the Stars* (New York: Penguin, 2017).

113 washed up at Cerro Ballena: R. Sagar and R. D. Cannon, "A Deep UBVRI CCD Photometric Study of the Moderately Young Southern Open Star Cluster NGC 4755 = K Crucis," *Astronomy and Astrophysics Supplement Series* 111 (1995): 75–84.

8. THE AGE OF GIANTS

117 two-story house: Stephen Leatherwood et al., *Whales, Dolphins, and Porpoises of the Eastern North Pacific and Adjacent Arctic Waters: A Guide to Their Identification* (New York: Dover, 1988).

117 enough to fill half a cement-mixer truck: The value is roughly a comparable value to the lung volume estimate (2,250 liters) for blue whales presented by Christina Lockyer, "Growth and Energy Budgets of Large Baleen Whales from the Southern Hemisphere," *Food and Agriculture Organization Fisheries Series* 3 (1981): 379–487.

117 diameter of dinner plates: An estimate based on comparable anatomical landmarks provided by Mai Nguyen, "How Scientists Preserved a 200 kg Blue Whale Heart," *Wired*, September 5, 2017, https://www.wired.com/story/how-scientists-preserved-a-440-pound-blue-whale-heart/.

117 twenty billion miles: A rough estimate derived from scaling isometrically the blood vessel length of the entire human body (average weight about sixty kg) at one hundred million miles to a blue whale. There are several sources for blood vessel length in the human body; see, for example, William C. Aird, "Spatial and Temporal Dynamics of the Endothelium," *Journal of Thrombosis and Haemostasis* 3 (2005): 1392–406.

117 **to every cell:** One thousand trillion is a quadrillion. It is an open question why blue whales, among other whales, do not have elevated cancer rates, at this scale of biomass. This question, called Peto's paradox, and the values for this figure can be found in Aleah F. Caulin et al., "Solutions to Peto's Paradox Revealed by Mathematical Modelling and Cross-Species Cancer Gene Analysis," *Philosophical Transactions of the Royal Society B* 370 (2015): 20140222, and references therein.

117 **over nine hundred miles:** See Ana Širović, John A. Hildebrand, and Sean M. Wiggins, "Blue and Fin Whale Call Source Levels and Propagation Range in the Southern Ocean," *Journal of the Acoustical Society of America* 122 (2007): 1208–15, and references therein.

118 **enshrined in textbooks:** One of the classic illustrations can be seen in E. J. Slijper, *Whales* (London: Hutchinson, 1962).

118 **extending in opposite directions:** Johan Nicolay Tønnessen and Arne Odd Johnsen, *The History of Modern Whaling* (Berkeley: University of California Press, 1982).

118 **relatively complete skeletons:** Mark Hallett and Mathew J. Wedel, *The Sauropod Dinosaurs: Life in the Age of Giants* (Baltimore: Johns Hopkins University Press, 2016).

118 **killed in 1926 by Norwegian whalers:** S. Risting, "Whales and Whale Foetuses: Statistics of Catch and Measurements Collected from the Norwegian Whalers' Association 1922–1925," *Rapports et Procès-Verbaux Des Réunions* 50 (1928): 1–122. Risting provided the most compelling evidence for the longest blue whale ever, and it has been cited explicitly as the ultimate length measurement, though the length value is sometimes mentioned without citation. Risting (p. 29) reported that the largest individual caught during 1922–1925 whaling seasons by the Norwegian Whalers' Association in the Southern Ocean was a female "106 feet [Norwegian feet] (33.27 metres) in length," killed in March 1926 near the South Shetland Islands. It is fair to assume that Risting reported the first value in Norwegian feet because this value was nearly correctly converted in the original publication to its metric value of 33.26 meters (or 109.11 feet). Risting further noted the rarity of individuals greater than 100 feet long (again in Norwegian feet, presumably), which consisted of only 5 individuals, out of 6,925 individuals killed from 1922 to 1925 by Norwegian operations in the Southern Ocean. For more about these data, see Trevor A. Branch et al., "Past and Present Distribution, Densities and Movements of Blue Whales *Balaenoptera musculus* in the Southern Hemisphere and Northern Indian Ocean," *Mammal Review* 37 (2007): 116–75.

118 just over 300,000 pounds: Waldon C. Winston, "The Largest Whale Ever Weighed," *Natural History Magazine*, 1950. It is clear that Winston's record of a 27.1-meter-long female blue whale (N. No. 319. B.F.) collected on January 27, 1948, by Japanese whalers in the Southern Ocean is the heaviest value among reliable measurements of blue whales. This account, described in particular detail, enumerates the piecemeal weighing of a 27.1-meter female blue whale with clear accounting for all of the individual tissue types (e.g., blubber, bone, organs, etc.), along with the specific limits of the individual tares and subset tabulations. Winston's final tally at 136.4 metric tons for this specimen explicitly did not include any correction for fluid loss, which might range from 6 percent to 10 percent of total body weight (see Lockyer, 1976).

119 about ten thousand times: Philip D. Gingerich, "Paleobiological Perspectives on Mesonychia, Archaeoceti, and the Origin of Whales," in *The Emergence of Whales* (New York: Springer, 1998), pp. 423–49.

119 essentially never recovered: Nicholas D. Pyenson and Simon N. Sponberg, "Reconstructing Body Size in Extinct Crown Cetacea (Neoceti) Using Allometry, Phylogenetic Methods and Tests from the Fossil Record," *Journal of Mammalian Evolution* 18 (2011): 269–88.

121 across different lineages: Graham J. Slater, Jeremy A. Goldbogen, and Nicholas D. Pyenson, "Independent Evolution of Baleen Whale Gigantism Linked to Plio-Pleistocene Ocean Dynamics," *Proceedings of the Royal Society B: Biological Sciences* 284 (2017): 20170546.

121 only a few million years ago: Robert W. Boessenecker, "Pleistocene Survival of an Archaic Dwarf Baleen Whale (Mysticeti: Cetotheriidae)," *Naturwissenschaften* 100 (2013): 365–71.

122 the worlds they lived in: Gene Hunt and Kaustuv Roy, "Climate Change, Body Size Evolution, and Cope's Rule in Deep-Sea Ostracodes," *Proceedings of the National Academy of Sciences* 103, no. 5 (2006): 1347–52; Daniel W. McShea, "Mechanisms of Large-Scale Evolutionary Trends," *Evolution* 48 (1994): 1747–63.

122 sucking it into their mouths: Alexander J. Werth, "Feeding in Marine Mammals," in *Feeding: Form, Function and Evolution in Tetrapod Vertebrates* (2000), pp. 475–514.

122 fingernails, hooves, and hair: Alexander J. Werth, "How Do Mysticetes Remove Prey Trapped in Baleen?" *Bulletin of the Museum of Comparative Zoology* 156 (2001): 189–203; Alexander J. Werth and Jean Potvin, "Baleen Hydrodynamics and Morphology of Cross-Flow Filtration in Balaenid Whale Suspension Feeding," *PLoS ONE* 11 (2016): e0150106.

123 **they once had teeth:** Thomas A. Deméré et al., "Morphological and Molecular Evidence for a Stepwise Evolutionary Transition from Teeth to Baleen in Mysticete Whales," *Systematic Biology* 57 (2008): 15–37.

123 **evidence that underlie this point:** In Darwin's first edition of *The Origin of Species*, he imagined how a population of black bears skimming insects in the water, if selected over long periods of time, could provide an analog for how filter feeding could evolve in aquatic animals, such as baleen whales. This idea was omitted in later editions, although by 1872 Darwin pointed to the lamellae in a duck's mouth, used as a kind of filter for feeding, as another useful analog for baleen in whales. See Carlos Peredo et al., "Decoupling Tooth Loss from the Evolution of Baleen in Whales," *Frontiers in Marine Science* 4 (2017): 67.

125 **dinosaurs with feathers:** Richard O. Prum and Andrew H. Brush, "Which Came First, the Feather or the Bird?" *Scientific American* 288 (2003): 84–93.

125 **primitive ridge of baleen:** Peredo et al., 2017.

126 **this part of the world:** Douglas Emlong was among the more prolific collectors. He collected what would turn out to be the first so-called toothed baleen whale, *Aetiocetus cotylalveus*. See Douglas Emlong, "A New Archaic Cetacean from the Oligocene of Northwest Oregon," *Bulletin of the Oregon University Museum of Natural History* 3 (1966): 1–51; and Clayton E. Ray, "Fossil Marine Mammals of Oregon," *Systematic Zoology* 25 (1976): 420–36. In recent decades, James L. Goedert and Kent Gibson, residents of Washington State and Oregon, respectively, have also donated to the Smithsonian important specimens that they have collected.

9. THE OCEAN'S UTMOST BONES

129 **some yet to be named:** Philip A. Morin et al., "Genetic Structure of the Beaked Whale Genus *Berardius* in the North Pacific, with Genetic Evidence for a New Species," *Marine Mammal Science* 33 (2017): 96–111.

129 **the catalog number:** The abbreviation "USNM" is a catalog shorthand for the United States National Museum, a bureaucratic entity that no longer technically exists at the Smithsonian yet served as the institution's foundational collection. The natural history collections that belonged to the United States National Museum in the nineteenth and early to mid twentieth centuries now form the National Museum of Natural History, or NMNH. Other parts of the USNM

collections, once housed in the Smithsonian's Arts and Industries Building and the Smithsonian's Castle, have been distributed to other units throughout the Smithsonian Institution. For brevity and convenience, my colleagues and I still refer to NMNH's collections as "the USNM," and the specimens that we collect for the National Collections still receive a USNM catalog number to this day.

129 **in any museum in the world:** Nicholas D. Pyenson, Jeremy A. Goldbogen, and Robert E. Shadwick, "Mandible Allometry in Extant and Fossil Balaenopteridae (Cetacea: Mammalia): The Largest Vertebrate Skeletal Element and Its Role in Rorqual Lunge Feeding," *Biological Journal of the Linnean Society* 108 (2013): 586–99.

130 **by the time we are born:** Tim D. White et al., *Human Osteology* (San Diego: Academic Press, 2011).

130 **to borrow from Melville:** Herman Melville, *Moby-Dick; or, The Whale* (New York: Harper and Brothers, 1851). The specific lines, from chapter 9 of *Moby-Dick*, pertain more to a description of depth of the seafloor: "Yet even then beyond the reach of any plummet—'out of the belly of hell'—when the whale grounded upon the ocean's utmost bones, even then, God heard the engulphed, repenting prophet when he cried."

131 **South America and Antarctica separated:** Graeme Eagles and Wilfried Jokat, "Tectonic Reconstructions for Paleobathymetry in Drake Passage," *Tectonophysics* 611 (2014): 28–50; and Howie D. Scher et al., "Onset of Antarctic Circumpolar Current 30 Million Years Ago as Tasmanian Gateway Aligned with Westerlies," *Nature* 523 (2015): 580–83.

132 **help: a whaling station:** The literature on Shackleton and his expeditions is extensive. My favorite retelling is a relatively recent one: Caroline Alexander, *The Endurance: Shackleton's Legendary Antarctic Expedition* (London: Bloomsbury, 1998). For the uninitiated, the original narratives are good starting points: Ernest H. Shackleton, *South: The Story of Shackleton's Last Expedition, 1914–1917* (New York: Macmillan, 1920); and Alfred Lansing, *Endurance: Shackleton's Incredible Voyage* (New York: Carroll and Graf Publishers, 1959).

132 **Norwegian, British, and Argentine:** R. Headland, *The Island of South Georgia* (Cambridge, UK: Cambridge University Press, 1984).

132 **process their catch on land:** Tønnessen and Johnsen, 1982.

132 **echoing throughout the harbor:** Tønnessen and Johnsen, 1982.

132 **whales killed last century:** Rocha Jr. et al., 2014.

133 a legacy of images: Frank Hurley, *South with Endurance: Shackleton's Antarctic Expedition 1914–1917: The Photographs of Frank Hurley* (New York: Simon & Schuster, 2001).

133 whaling in the twentieth century: Trevor A. Branch et al., "Past and Present Distribution, Densities and Movements of Blue Whales *Balaenoptera musculus* in the Southern Hemisphere and Northern Indian Ocean," *Mammal Review* 37 (2007): 116–75.

133 lengths of their ancestors: Jennifer A. Jackson et al., "How Few Whales Were There After Whaling? Inference from Contemporary mtDNA Diversity," *Molecular Ecology* 17 (2008): 236–51.

134 form of oil and meat: D. Graham Burnett, *The Sounding of the Whale: Science and Cetaceans in the Twentieth Century* (Chicago: University of Chicago Press, 2012).

134 catch statistics for decades: Alexey V. Yablokov, "Validity of Whaling Data," *Nature* 367 (1994): 108; Alfred A. Berzin, "The Truth About Soviet Whaling," *Marine Fisheries Review* 70 (2008): 4–59; Yulia V. Ivashchenko and Phillip J. Clapham, "Too Much Is Never Enough: The Cautionary Tale of Soviet Illegal Whaling," *Marine Fisheries Review* 76 (2014): 1–22.

134 from the whaling industry: N. A. Mackintosh and J. F. G. Wheeler, "Southern Blue and Fin Whales," *Discovery Reports* 1 (1929): 257–540.

135 at the Smithsonian, Remington Kellogg: Burnett, 2012. One example of Kellogg's legacy is the continuing relevance of the IWC's annual Scientific Committee. The annual meeting of the committee is one of the premier venues for whale scientists across the world to share the most current data on the status of cetacean populations. The exchange includes a range of topics, from genetics and ecology to the impact of different kinds of pollution and, of course, whaling.

136 many fisheries do today: Stan Ulanski, *The Billfish Story: Swordfish, Sailfish, Marlin, and Other Gladiators of the Sea* (Athens: University of Georgia Press, 2013); Paul Greenberg, *Four Fish: The Future of the Last Wild Food* (New York: Penguin Press, 2010).

136 still poorly understood: James A. Estes et al., eds., *Whales, Whaling, and Ocean Ecosystems* (Berkeley: University of California Press, 2007); James A. Estes et al., "Trophic Downgrading of Planet Earth," *Science* 333 (2011): 301–6.

137 about the specimen collection: Quentin R. Walsh and P. J. Capelotti, *The Whaling Expedition of the Ulysses, 1937–38* (Gainesville: University Press of Florida, 2010).

140 a mechanical advantage: Pyenson et al., 2013.
141 hunt-and-peck accounting: Jeremy A. Goldbogen, Jean Potvin, and Robert E. Shadwick, "Skull and Buccal Cavity Allometry Increase Mass-Specific Engulfment Capacity in Fin Whales," *Proceedings of the Royal Society of London B: Biological Sciences* 277 (2010): 861–68.

10. A DISCOVERY AT HVALFJÖRÐUR

143 uses its wing flaps: Frank E. Fish and George V. Lauder, "Control Surfaces of Aquatic Vertebrates: Active and Passive Design and Function," *Journal of Experimental Biology* 220 (2017): 4351–63.
144 the machinelike processes: Steven Vogel, *Life's Devices: The Physical World of Animals and Plants* (Princeton, NJ: Princeton University Press, 1988).
144 one species (fin whales): Jeremy A. Goldbogen, Nicholas D. Pyenson, and Robert E. Shadwick, "Big Gulps Require High Drag for Fin Whale Lunge Feeding," *Marine Ecology Progress Series* 349 (2007): 289–301.
145 farm tractor tire: Dimensions estimated from Mai Nguyen, "How Scientists Preserved a 440-Pound Blue Whale Heart," *Wired*, July 2, 2017, www.wired.com/story/how-scientists-preserved-a-440-pound -blue-whale-heart/.
145 thousands of liters of blood: John M. Gosline and Robert E. Shadwick, "The Mechanical Properties of Fin Whale Arteries Are Explained by Novel Connective Tissue Designs," *Journal of Experimental Biology* 199 (1996): 985–97.
145 every cycle of the heart pumping: Drake et al., eds., *Gray's Anatomy for Students* (Philadelphia: Churchill Livingstone/Elsevier, 2014).
145 Bob and his colleagues discovered: Gosline and Shadwick, 1996.
146 beating heart of a wild baleen whale: Eric A. Wahrenbrock et al., "Respiration and Metabolism in Two Baleen Whale Calves," *Marine Fisheries Review* 36 (1974): 3–8.
146 but never farther: Paul F. Brodie, "Feeding Mechanics of Rorquals *Balaenoptera* sp.," in Jean-Michel Mazin and Vivian de Buffrénil, eds., *Secondary Adaptation of Tetrapods to Life in Water: Proceedings of the International Meeting, Poitiers, 1996* (Munich: Verlag Dr. Friedrich Pfeil, 2001).
147 roughly 50,000 individuals: Stephen B. Reilly et al., "*Balaenoptera physalus*," *The IUCN Red List of Threatened Species* (2013):

e.T2478A44210520, http://dx.doi.org/10.2305/IUCN.UK.2013-1.RLTS
.T2478A44210520.en.

151 over a century old: Among many others, see H. von W. Schulte, "Anatomy of a Foetus of *Balaenoptera borealis*," *Memoirs of the American Museum of Natural History*, New Series 1, part 6 (1916): 389–502.

152 How quickly does it inflate: Marina A. Piscitelli et al., "Lung Size and Thoracic Morphology in Shallow- and Deep-Diving Cetaceans," *Journal of Morphology* 271 (2010): 654–73.

152 how long they could feed: A. Acevedo-Gutierrez, D. A. Croll, and B. R. Tershy, "High Feeding Costs Limit Dive Time in the Largest Whales," *Journal of Experimental Biology* 205 (2002): 1747–53.

152 lodged in the neck vertebrae: M. A. Lillie et al., "Cardiovascular Design in Fin Whales: High-Stiffness Arteries Protect Against Adverse Pressure Gradients at Depth," *Journal of Experimental Biology* 216 (2013): 2548–63.

152 countercurrent exchange system: Per Fredrik Scholander and William E. Schevill, "Counter-Current Vascular Heat Exchange in the Fins of Whales," *Journal of Applied Physiology* 8 (1955): 279–82; Knut Schmidt-Nielsen, "Countercurrent Systems in Animals," *Scientific American* 244 (1981): 118–29.

152 part of the fin whale's body: John E. Heyning and James G. Mead, "Thermoregulation in the Mouths of Feeding Gray Whales," *Science* 278 (1997): 1138–40.

154 the course of a single lunge: Brodie, 2001.

155 your own ignorance: This idea has been around for some time. Daniel J. Boorstin, in an interview, probably articulated it the best: "The greatest obstacle to discovery is not ignorance—it is the illusion of knowledge." Carol Krucoff, "The 6 O'Clock Scholar," *Washington Post*, January 29, 1984, https://www.washingtonpost.com/archive/lifestyle/1984/01/29/the-6-oclock-scholar.

156 scoured the anatomical reprints: Including, among many, Schulte, 1916. We also consulted August Pivorunas, "The Fibrocartilage Skeleton and Related Structures of the Ventral Pouch of Balaenopterid Whales," *Journal of Morphology* 151 (1977): 299–313, and references therein.

11. PHYSICS AND FLENSING KNIVES

157 feed like a rorqual whale: J. A. Goldbogen et al., "How Baleen Whales Feed: The Biomechanics of Engulfment and Filtration," *Annual Review of Marine Science* 9 (2017): 367–86.

157 swallow your meal: Alexander J. Werth and Haruka Ito, "Sling, Scoop, and Squirter: Anatomical Features Facilitating Prey Transport, Processing, and Swallowing in Rorqual Whales (Mammalia: Balaenopteridae)," *Anatomical Record* 300 (2017): 2070–86; R. H. Lambertsen, "Internal Mechanism of Rorqual Feeding," *Journal of Mammalogy* 64 (1983): 76–88.

158 surface of the throat pouch: Lisa S. Orton and Paul F. Brodie, "Engulfing Mechanics of Fin Whales," *Canadian Journal of Zoology* 65 (1987): 2898–907.

159 to the rorqual throat pouch: Jean Potvin, Jeremy A. Goldbogen, and Robert E. Shadwick, "Scaling of Lunge Feeding in Rorqual Whales: An Integrated Model of Engulfment Duration," *Journal of Theoretical Biology* 267 (2010): 437–53.

159 worked at different sizes: Jean Potvin, Jeremy A. Goldbogen, and Robert E. Shadwick, "Metabolic Expenditures of Lunge Feeding Rorquals Across Scale: Implications for the Evolution of Filter Feeding and the Limits to Maximum Body Size," *PLoS ONE* 7 (2012): e44854.

159 active throat pouch expansion: Jean Potvin, Jeremy A. Goldbogen, and Robert E. Shadwick, "Passive Versus Active Engulfment: Verdict from Trajectory Simulations of Lunge-Feeding Fin Whales," *Journal of the Royal Society Interface* 6 (2009): 1005–25.

161 embedded in the blubber layer: Pivorunas, 1977.

161 nerves played a roll in lunge feeding: A. Wayne Vogl et al., "Stretchy Nerves: Essential Components of an Extreme Feeding Mechanism in Rorqual Whales," *Current Biology* 25 (2015): R360–61.

164 appropriate permits for transit: All tissue samples from Iceland were transferred and imported to Canada under permits granted by the Convention on International Trade in Endangered Species of Wild Fauna and Flora.

165 trapped in the chin of a whale: Nicholas D. Pyenson et al., "Discovery of a Sensory Organ That Coordinates Lunge Feeding in Rorqual Whales," *Nature* 485 (2012): 498–501.

166 exclusively on one side: Also, the fin whale is the only mammal on the planet that is consistently asymmetrically pigmented, with a blaze of white on its right throat and jaw, but black on its left. Bernie R. Tershy and David N. Wiley, "Asymmetrical Pigmentation in the Fin Whale: A Test of Two Feeding Related Hypotheses," *Marine Mammal Science* 8 (1992): 315–18.

12. THE LIMITS OF LIVING THINGS

170 get big over geologic time: John Alroy, "Cope's Rule and the Dynamics of Body Mass Evolution in North American Fossil Mammals," *Science* 280 (1998): 731–34.

170 tens of thousands of pounds: Roger B. J. Benson et al., "Cope's Rule and the Adaptive Landscape of Dinosaur Body Size Evolution," *Palaeontology* 61 (2017): 13–48.

170 about the same amount of time: Felisa A. Smith et al., "The Evolution of Maximum Body Size of Terrestrial Mammals," *Science* 330 (2010): 1216–19.

170 sea cows to evolve: Nicholas D. Pyenson and Geerat J. Vermeij, "The Rise of Ocean Giants: Maximum Body Size in Cenozoic Marine Mammals as an Indicator for Productivity in the Pacific and Atlantic Oceans," *Biology Letters* 12 (2016): 20160186.

171 inside or outside the organism: John T. Bonner, *Why Size Matters: From Bacteria to Blue Whales* (Princeton, NJ: Princeton University Press, 2011); Knut Schmidt-Nielsen, *Scaling: Why Is Animal Size So Important?* (Cambridge, UK: Cambridge University Press, 1984).

172 the cost of overheating: Peter J. Corkeron and Richard C. Connor, "Why Do Baleen Whales Migrate?" *Marine Mammal Science* 15 (1999): 1228–45.

172 cetaceans appear to obey: Jerry F. Downhower and Lawrence S. Blumer, "Calculating Just How Small a Whale Can Be," *Nature* 335 (1988): 675.

172–73 and live longer: Geoffrey B. West and James H. Brown, "Life's Universal Scaling Laws," *Physics Today* 57 (2004): 36–43.

173 surface area to its volume: West and Brown, 2004.

173 at the surface afterward: Marina A. Piscitelli et al., "A Review of Cetacean Lung Morphology and Mechanics," *Journal of Morphology* 274 (2013): 1425–40.

173 with oxygen-storing enhancements: Shawn R. Noren and Terrie M. Williams, "Body Size and Skeletal Muscle Myoglobin of Cetaceans: Adaptations for Maximizing Dive Duration," *Comparative Biochemistry and Physiology Part A: Molecular & Integrative Physiology* 126 (2000): 181–91.

174 such as whales and seals: Scott Mirceta et al., "Evolution of Mammalian Diving Capacity Traced by Myoglobin Net Surface Charge," *Science* 340 (2013): 1234192.

174 **as in birds today:** Mathew J. Wedel, "A Monument of Inefficiency: The Presumed Course of the Recurrent Laryngeal Nerve in Sauropod Dinosaurs," *Acta Palaeontologica Polonica* 57 (2011): 251–56.

174 **living life as a blue whale:** Potvin, Goldbogen, and Shadwick, 2012.

176 **we still don't really know:** M. M. Walker et al., "Evidence That Fin Whales Respond to the Geomagnetic Field During Migration," *Journal of Experimental Biology* 171 (1992): 67–78.

176 **together using bubble nets:** David Wiley et al., "Underwater Components of Humpback Whale Bubble-Net Feeding Behaviour," *Behaviour* 148 (2011): 575–602.

176 **assemble and disassemble randomly:** Jenny Allen et al., "Network-Based Diffusion Analysis Reveals Cultural Transmission of Lobtail Feeding in Humpback Whales," *Science* 340 (2013): 485–88.

177 **kind of humpback culture:** Hal Whitehead and Luke Rendell, *The Cultural Lives of Whales and Dolphins* (Chicago: University of Chicago Press, 2014).

180 **its orbit around the Sun:** Jamie Woodward, *The Ice Age: A Very Short Introduction*, vol. 380 (Oxford, UK: Oxford University Press, 2014).

180 **shorter amounts of time:** J. R. Marlow et al., "Upwelling Intensification as Part of the Pliocene-Pleistocene Climate Transition," *Science* 290 (2000): 2288–91.

181 **confers the same advantages:** Goldbogen et al., 2017.

13. ARCTIC TIME MACHINES

185 **on a course for disaster:** The literature on the Franklin expedition continues to grow, especially now that the HMS *Terror* and *Erebus* have been located in the Canadian Arctic. See Paul Watson, *Ice Ghosts: The Epic Hunt for the Lost Franklin Expedition* (New York: W. W. Norton, 2017).

186 **missions meant to find them:** W. Gillies Ross, "The Type and Number of Expeditions in the Franklin Search 1847–1859," *Arctic* 55 (2002): 57–69.

186 **I imagine Franklin:** Franklin's first wife had died during his previous expeditions to the north water, leaving an infant daughter without a father at home. At more than sixty years old in 1835, Franklin was not Britain's first choice to command this expedition, and he burned with the need to vindicate his name after serving as a civilian

administrator in Tasmania. See Kathleen Fitzpatrick, "Franklin, Sir John (1786–1847)," in Douglas Pike, ed., *Australian Dictionary of Biography, Vol. 1, 1788–1850, A–H* (Melbourne, Australia: Melbourne University Publishing, 1966), available at http://adb.anu.edu.au/biography/franklin-sir-john-2066/text2575.

186 **side of the *Erebus*:** It is worth pointing out that both the *Erebus* and *Terror* had just returned from a mission to Antarctica from 1839 to 1843 before heading to the Arctic in 1845. Under the leadership of Sir James Clark Ross, the purpose of the Ross expedition was primarily scientific exploration, and their findings generated many publications, including the first account of the Ross seal, which remains one of the least well known (and observed) of the Antarctic seals. See M. J. Ross, *Ross in the Antarctic: The Voyages of James Clark Ross in Her Majesty's Ships Erebus and Terror, 1839–1843* (Whitby, UK: Caedmon of Whitby, 1982).

188 **around a breathing hole:** John C. George et al., "Observations on the Ice-Breaking and Ice Navigation Behavior of Migrating Bowhead Whales (*Balaena mysticetus*) Near Point Barrow, Alaska, Spring 1985," *Arctic* 42 (1989): 24–30.

188 **under these circumstances:** Morten P. Porsild, "On 'Savssats': A Crowding of Arctic Animals at Holes in the Sea Ice," *Geographical Review* 6 (1918): 215–28; Mats P. Heide-Jørgensen et al., "Three Recent Ice Entrapments of Arctic Cetaceans in West Greenland and the Eastern Canadian High Arctic," *NAMMCO Scientific Publications* 4 (2002): 143–48.

188 **rest on the ocean floor:** Watson, 2017.

188 **down from the North Pole:** Sue E. Moore and Kristin L. Laidre, "Trends in Sea Ice Cover Within Habitats Used by Bowhead Whales in the Western Arctic," *Ecological Applications* 16 (2006): 932–44.

189 **pounds of growth per day:** Christina Lockyer, "Review of Baleen Whale (Mysticeti) Reproduction and Implications for Management," *Reports of the International Whaling Commission, Special Issue 6* (1984): 27–50.

189 **matured more slowly:** John C. George et al., "Age and Growth Estimates of Bowhead Whales (*Balaena mysticetus*) via Aspartic Acid Racemization," *Canadian Journal of Zoology* 77 (1999): 571–80.

189 **not trivial or straightforward:** Aleta A. Hohn, "Age Estimation," in W. F. Perrin et al., eds., *Encyclopedia of Marine Mammals*, 2nd ed. (San Diego, CA: Academic Press, 2009), pp. 11–17.

189 **like tree rings:** W. F. Perrin and A. C. Myrick Jr., eds., "Age Determination of Toothed Whales and Sirenians," *Reports of the International Whaling Commission, Special Issue 3* (1980): 1–229.

190 **certainly not a lifetime:** S. C. Lubetkin et al., "Age Estimation for Young Bowhead Whales (*Balaena mysticetus*) Using Annual Baleen Growth Increments," *Canadian Journal of Zoology* 86 (2008): 525–38.

190 **layers to be read:** Cheryl Rosa et al., "Age Estimates Based on Aspartic Acid Racemization for Bowhead Whales (*Balaena mysticetus*) Harvested in 1998–2000 and the Relationship Between Racemization Rate and Body Temperature," *Marine Mammal Science* 29 (2013): 424–45.

190 **Alaska was still Russian territory:** John Murdoch, *Ethnological Results of the Point Barrow Expedition*, Ninth Annual Report of the Bureau of Ethnology to the Secretary of the Smithsonian 1887–88. (Washington, DC: Government Printing Office, 1892).

190 **the nineteenth century: 133:** John C. George et al., "A New Way to Estimate the Age of Bowhead Whales (*Balaena mysticetus*) Using Ovarian Corpora Counts," *Canadian Journal of Zoology* 89 (2011): 840–52.

191 **a process called racemization:** J. L. Bada and S. E. Brown, "Amino Acid Racemization in Living Mammals: Biochronological Applications," *Trends in Biochemical Sciences* 5 (1980): 3–5.

191 **astonishing 211 years old:** For this individual whale, 95WW5, the standard error for its age estimate was 35 years, yielding an upper bound of 246 years or a lower bound of 176 years. Craig and his colleagues have reported ten bowheads with estimated ages greater than 100 years, including an additional animal, 95B9, with an upper age bound of 201 years. See George et al., 1999 and 2011.

192 **to form a dense mat:** Alexander J. Werth, "Models of Hydrodynamic Flow in the Bowhead Whale Filter Feeding Apparatus," *Journal of Experimental Biology* 207 (2004): 3569–80.

193 **an Arctic time machine:** Sang Heon Lee et al., "Regional and Seasonal Feeding by Bowhead Whales *Balaena mysticetus* as Indicated by Stable Isotope Ratios," *Marine Ecology Progress Series* 285 (2005): 271–87.

194 **food web in the Arctic:** Paul Szpak et al., "Long-Term Ecological Changes in Marine Mammals Driven by Recent Warming in Northwestern Alaska," *Global Change Biology* 24 (2018): 490–503.

195 **threat to right whales:** North Atlantic right whales remain in serious peril. In 2017, seventeen whales were killed, with causes mostly attributable to the trauma of ship strike or fishing net entanglement. This rate of mortality is a major threat for a species consisting of about 450 individual whales. See Stephanie Taylor and Tony R. Walker, "North Atlantic Right Whales in Danger," *Science* 358 (2017): 730–31.

196 **lack of effort:** Paul Wade et al., "Acoustic Detection Satellite-Tracking Leads to Discovery of Rare Concentration of Endangered North Pacific Right Whales," *Biology Letters* 2 (2006): 417–19.

196 **faster than the whales can swim:** Scott D. Kraus et al., "North Atlantic Right Whales in Crisis," *Science* 309 (2005): 561–62.

196 **more frequent and persistent:** Jeff W. Higdon and Steven H. Ferguson, "Loss of Arctic Sea Ice Causing Punctuated Change in Sightings of Killer Whales (*Orcinus orca*) over the Past Century," *Ecological Applications* 19 (2009): 1365–75.

196 **on bowhead fins and flukes:** J. Craig George et al., "Frequency of Injuries from Line Entanglements, Killer Whales, and Ship Strikes on Bering-Chukchi-Beaufort Seas Bowhead Whales," *Arctic* 70 (2017): 37–46.

197 **base of Arctic food webs:** Kristin L. Laidre et al., "Arctic Marine Mammal Population Status, Sea Ice Habitat Loss, and Conservation Recommendations for the 21st Century," *Conservation Biology* 29 (2015): 724–37.

198 **suddenly and with little warning:** Anthony D. Barnosky et al., "Approaching a State Shift in Earth's Biosphere," *Nature* 486 (2012): 52–58.

198 **the Canadian Arctic Archipelago:** Arthur S. Dyke, James Hooper, and James M. Savelle, "A History of Sea Ice in the Canadian Arctic Archipelago Based on Postglacial Remains of the Bowhead Whale (*Balaena mysticetus*)," *Arctic* 49 (1996): 235–55.

199 **both material and acoustic:** And especially plastics, which are pervasive and break down but persist in foodwebs as tiny particles. Elitza S. Germanov et al., " Microplastics: No Small Problem for Filter-Feeding Megafauna," *Trends in Ecology & Evolution* (2018): 227–32.

199 **as the Arctic unravels:** Data and facts mentioned in this chapter derive primarily from the Arctic Council's 2017 SWIPA report, and reference therein. Arctic Monitoring and Assessment Programme, *Snow, Water, Ice and Permafrost: Summary for Policy-makers* (Oslo: AMAP, 2017).

14. SHIFTING BASELINES

200 ideas underpinning ecology: Jonathan M. Chase, "Are There Real Differences Among Aquatic and Terrestrial Food Webs?" *Trends in Ecology & Evolution* 15 (2000): 408–12.

202 directly on the sun harvesters: Andrew W. Trites, "Food Webs in the Ocean: Who Eats Whom and How Much," in Michael Sinclair and Grimur Valdimarsson, eds., *Responsible Fisheries in the Marine Ecosystem* (Wallingford, UK: CABI Publishing/Food and Agriculture Organization of the United Nations, 2003), pp. 125–41.

202 biosphere's top consumer: Estes et al., 2011.

202 inverted at the base: Jonathan B. Shurin, Daniel S. Gruner, and Helmut Hillebrand, "All Wet or Dried Up? Real Differences Between Aquatic and Terrestrial Food Webs," *Proceedings of the Royal Society of London B: Biological Sciences* 273 (2006): 1–9.

203 exclusivity of the bottom-up view: Estes et al., eds., 2007.

203 fond of eating sea urchins: J. A. Estes and J. F. Palmisano, "Sea Otters: Their Role in Structuring Nearshore Communities," *Science* 185 (1974): 1058–60.

203 Estes and his colleagues argued: James A. Estes et al., "Killer Whale Predation on Sea Otters Linking Oceanic and Nearshore Ecosystems," *Science* 282 (1998): 473–76.

204 how the world once was: Jeremy B. C. Jackson, "Ecological Extinction and Evolution in the Brave New Ocean," *Proceedings of the National Academy of Sciences* 105 (2008): 11458–65.

205 populations was once like: Daniel Pauly, "Anecdotes and the Shifting Baseline Syndrome of Fisheries," *Trends in Ecology & Evolution* 10 (1995): 430.

205 than they are today: Joe Roman and Stephen R. Palumbi, "Whales Before Whaling in the North Atlantic," *Science* 301 (2003): 508–10.

206 to the ocean ecosystem: Joe Roman and James J. McCarthy, "The Whale Pump: Marine Mammals Enhance Primary Productivity in a Coastal Basin," *PLoS ONE* 5 (2010): e13255.

15. ALL THE WAYS TO GO EXTINCT

210–11 in recent Earth history: Aaron O'Dea et al., "Formation of the Isthmus of Panama," *Science Advances* 2 (2016): e1600883.

211 **separated evolutionary paths:** Egbert G. Leigh, Aaron O'Dea, and Geerat J. Vermeij, "Historical Biogeography of the Isthmus of Panama," *Biological Reviews* 89 (2014): 148–72.

215 **the last survey in 2006:** Samuel T. Turvey et al., "First Human-Caused Extinction of a Cetacean Species?" *Biology Letters* 3 (2007): 537–40.

215 **we can read Hoy's account:** C. M. Hoy, " The 'White-Flag Dolphin' of the Tung Ting Lake," *China Journal of Arts and Science* 1 (1923): 154–57.

218 **to the Smithsonian in 1918:** G. S. Miller, "A New River Dolphin from China," *Smithsonian Miscellaneous Collections* 68 (1918): 1–12.

219 **In the midtwentieth century:** G. G. Simpson, "The Principles of Classification, and a Classification of Mammals," *Bulletin of the American Museum of Natural History* 85 (1945): 1–350.

219 **they had separate ancestries:** H. Hamilton et al., "Evolution of River Dolphins," *Proceedings of the Royal Society of London B: Biological Sciences* 268 (2001): 549–56.

220 **we christened *Isthminia panamensis*:** N. D. Pyenson et al., "*Isthminia panamensis*, a New Fossil Inioid (Mammalia, Cetacea) from the Chagres Formation of Panama and the Evolution of 'River Dolphins' in the Americas," *PeerJ* 3 (2015): e1227.

221 **"Bycatch" sounds abstract:** Bycatch kills an estimated 600,000 cetaceans every year, ranging from small to large species, globally. See Andrew J. Read, Phebe Drinker, and Simon Northridge, "Bycatch of Marine Mammals in US and Global Fisheries," *Conservation Biology* 20 (2006): 163–69.

221 **science only since 1958:** K. S. Norris and W. N. McFarland, "A New Harbor Porpoise of the Genus *Phocoena* from the Gulf of California," *Journal of Mammalogy* 39 (1958): 22–39.

221 **have looked for them:** A committee called CIRVA (International Committee for the Recovery of the Vaquita), formed by the U.S. Marine Mammal Commission, a federal watchdog agency, and the International Union for Conservation of Nature and Natural Resources, compilers of the Red List, has specifically been dealing with vaquita issues. CIRVA-9 (May 2017) is the latest document reporting on vaquitas' status, although most publicly available recent survey estimates are based on data from CIRVA-8 (November 2016).

222 **local and a world away:** Environmental Investigation Agency, "Dual Extinction: The Illegal Trade in the Endangered Totoaba and Its Impact on the Critically Endangered Vaquita," briefing to the 66th Standing Committee of CITES, January 11–15, 2016, https://drive

.google.com/viewerng/viewer?url=https://eia-international.org/wp-content/uploads/EIA-Dual-Extinction-mr.pdf.

223 **to corral the vaquita:** Nick Pyenson, "Ballad of the Last Porpoise," *Smithsonian*, November 2017, pp. 29–33.

223 **the stress of its capture:** Livia Albeck-Ripka, "30 Vaquita Porpoises Are Left; One Died in a Rescue Mission," *New York Times*, November 11, 2017.

224 **we might be wrong:** Samuel T. Turvey et al., "Spatial and Temporal Extinction Dynamics in a Freshwater Cetacean," *Proceedings of the Royal Society of London B: Biological Sciences* 277 (2010): 3139–47.

16. EVOLUTION IN THE ANTHROPOCENE

228 **as weapons in self-defense:** One of my favorite scientific papers describes the ecological framework of this evolutionary arms race within cetaceans: John K. B. Ford and Randall R. Reeves. "Fight or Flight: Antipredator Strategies of Baleen Whales," *Mammal Review* 38 (2008): 50–86.

228 **they eat sharks:** John K. B. Ford et al., "Shark Predation and Tooth Wear in a Population of Northeastern Pacific Killer Whales," *Aquatic Biology* 11 (2011): 213–24.

229 **tantalizing fossils nearby:** Alexandra T. Boersma and Nicholas D. Pyenson, "*Arktocara yakataga*, a New Fossil Odontocete (Mammalia, Cetacea) from the Oligocene of Alaska and the Antiquity of Platanistoidea," *PeerJ* 4 (2016): e2321.

230 **consisted of a harbor seal:** Stephen Raverty, personal communication, August 2017.

230 **In a follow-up:** Alan M. Springer et al., "Sequential Megafaunal Collapse in the North Pacific Ocean: An Ongoing Legacy of Industrial Whaling?," *Proceedings of the National Academy of Sciences of the USA* 100 (2003): 12223–28.

231 **smaller than they are:** Ford and Reeves, 2008.

231 **discomfiting to human onlookers:** John K. B. Ford, Graeme M. Ellis, and Kenneth C. Balcomb, *Killer Whales: The Natural History and Genealogy of Orcinus orca in British Columbia and Washington* (Vancouver: UBC Press, 2000).

232 **live into their nineties:** Lauren J. N. Brent et al., "Ecological Knowledge, Leadership, and the Evolution of Menopause in Killer Whales," *Current Biology* 25 (2015): 746–50.

232 **humanity's chemical legacy:** Robert C. Lacy et al., "Evaluating Anthropogenic Threats to Endangered Killer Whales to Inform Effective Recovery Plans," *Scientific Reports* 7 (2017): 14119.

232 **during lean times:** Larry Pynn, "The Hunger Games: Two Killer Whales, Same Sea, Different Diets," *Hakai Magazine*, November 28, 2017, https://www.hakaimagazine.com/features/hunger-games-two-killer-whales-same-sea-different-diets/.

232–33 **encephalization quotient, or EQ:** Hal Jerison, *Evolution of the Brain and Intelligence* (New York: Academic Press, 1973).

233 **but ahead of chimpanzees:** Lori Marino, Daniel W. McShea, and Mark D. Uhen, "Origin and Evolution of Large Brains in Toothed Whales," *Anatomical Record* 281 (2004): 1247–55.

233 **much like our own:** Lori Marino et al., "Cetaceans Have Complex Brains for Complex Cognition," *PLoS Biology* 5 (2007): e139.

233 **sophisticated and intelligent to us:** Lori Marino, "Convergence of Complex Cognitive Abilities in Cetaceans and Primates," *Brain, Behavior and Evolution* 59 (2002): 21–32.

234 **and, maybe, killer whales:** Diana Reiss and Lori Marino, "Mirror Self-recognition in the Bottlenose Dolphin: A Case of Cognitive Convergence," *Proceedings of the National Academy of Sciences* 98 (2001): 5937–42.

235 **chirps of whalesong:** Michael J. Noad et al., "Cultural Revolution in Whale Songs," *Nature* 408 (2000): 537.

235 **to minke whales:** Shannon Rankin and Jay Barlow, "Source of the North Pacific 'Boing' Sound Attributed to Minke Whales," *Journal of the Acoustical Society of America* 118 (2005): 3346–51.

235 **share among themselves:** M. Bearzi and C. B. Stanford, *Beautiful Minds: The Parallel Lives of Great Apes and Dolphins* (Cambridge, MA: Harvard University Press, 2008).

236 **whales possess culture:** Hal Whitehead and Luke Rendell, *The Cultural Lives of Whales and Dolphins* (Chicago: University of Chicago Press, 2014).

236 **distinct acoustic clans:** Luke E. Rendell and Hal Whitehead, "Vocal Clans in Sperm Whales (*Physeter macrocephalus*)," *Proceedings of the Royal Society of London B: Biological Sciences* 270 (2003): 225–31.

237 **cultural traditions with it:** Shane Gero, "The Lost Culture of Whales," *New York Times*, October 8, 2016.

237 **just the right size:** Ana D. Davidson et al., "Multiple Ecological Pathways to Extinction in Mammals," *Proceedings of the National Academy of Sciences* 106 (2009): 10702–5.

237 **urbanized ocean habitat:** Douglas J. McCauley et al., "Marine Defaunation: Animal Loss in the Global Ocean," *Science* 347 (2015): 1255641.

239 **Third, stay global:** Ana D. Davidson et al., "Drivers and Hotspots of Extinction Risk in Marine Mammals," *Proceedings of the National Academy of Sciences* 109 (2012): 3395–400. Based on the fossil record, broad geographic ranges definitely help species survive major extinction events. Jonathan L. Payne and Seth Finnegan, "The Effect of Geographic Range on Extinction Risk During Background and Mass Extinction," *Proceedings of the National Academy of Sciences* 104 (2007): 10506–11.

240 **true lineage splitting:** Transient and resident killer whale ecotypes in the North Pacific have been separate evolutionary units for the past 200,000 years, while Antarctic ecotypes have more recent divergences. See Andrew D. Foote et al., "Genome-Culture Coevolution Promotes Rapid Divergence of Killer Whale Ecotypes," *Nature Communications* 7 (2016): 11693.

17. WHALEBONE JUNCTION

242 **historical black-and-white image:** The historical images showing children by a whale skeleton at Whalebone Junction are archived at the Outer Banks History Center in Manteo, North Carolina, which is administered by the Special Collections Section of the State Archives of North Carolina. Also, see Michelle Wagner, "Whalebone Junction: Crossroads of the Outer Banks," *North Beach Sun*, February 15, 2017, http://www.northbeachsun.com/whalebone-junction -crossroads-of-the-outer-banks/.

243 **heart-shaped blows:** Leatherwood et al., 1988.

244 **remain poorly studied:** S. B. Reilly et al., *"Eschrichtius robustus," The IUCN Red List of Threatened Species* (2008): e.T8097A12885255, http://dx.doi.org/10.2305/IUCN.UK.2008.RLTS.T8097A12885255.en.

245 **Bering and Chukchi seas:** Joe Roman et al., "Whales as Marine Ecosystem Engineers," *Frontiers in Ecology and the Environment* 12 (2014): 377–85.

245 **grounds in Baja California:** Peter J. Bryant, Christopher M. Lafferty, and Susan K. Lafferty, "Reoccupation of Laguna Guerrero Negro, Baja California, Mexico, by Gray Whales," in Mary Lou Jones et al., eds., *The Gray Whale*: Eschrichtius robustus (Orlando, FL: Academic Press, 1984), pp. 375–87.

245 less than a hundred years: P. J. Clapham, S. B. Young, and R. L. Brownell, "Baleen Whales: Conservation Issues and the Status of the Most Endangered Populations," *Mammal Review* 29 (1999): 37–62.

246 both young animals: S. E. Noakes, N. D. Pyenson, and G. McFall, "Late Pleistocene Gray Whales (*Eschrichtius robustus*) Offshore Georgia, U.S.A., and the Antiquity of Gray Whale Migration in the North Atlantic Ocean," *Palaeogeography, Palaeoclimatology, Palaeoecology* 392 (2013): 502–9.

247 occupation across the North Atlantic: Ole Lindquist, *The North Atlantic Gray Whale (*Escherichtius [sic] robustus*): An Historical Outline Based on Icelandic, Danish-Icelandic, English and Swedish Sources Dating from ca 1000 AD to 1792*, Occasional papers 1 (St. Andrews, UK: University of St. Andrews, 2000).

247 stored at the Smithsonian: James G. Mead and Edward D. Mitchell, "Atlantic Gray Whales," in Jones et al., eds., *The Gray Whale*, pp. 33–53.

247 until about 450 years ago: Peter J. Bryant, "Dating Remains of Gray Whales from the Eastern North Atlantic," *Journal of Mammalogy* 76 (1995): 857–61.

247 portals across oceans: S. Elizabeth Alter, Eric Rynes, and Stephen R. Palumbi, "DNA Evidence for Historic Population Size and Past Ecosystem Impacts of Gray Whales," *Proceedings of the National Academy of Sciences* 104 (2007): 15162–67.

248 the most likely pathway: Aviad P. Scheinin et al., "Gray Whale (*Eschrichtius robustus*) in the Mediterranean Sea: Anomalous Event or Early Sign of Climate-Driven Distribution Change?" *Marine Biodiversity Records* 4 (2011): e28.

249 In 1883 one of my: Frederick W. True, "Suggestions to Keepers of the U.S. Lifesaving Stations, Light-houses, and Light-ships, and to Other Observers, Relative to the Best Means of Collection and Preserving Specimens of Whales and Porpoises," *Annual Reports of the United States Commission of Fish and Fisheries* 11, app. F (1883): 1157–82.

250 in Nags Head and Corolla: Mead and Mitchell, 1984.

251 to paraphrase Faulker: William Faulkner, *Requiem for a Nun* (New York: Random House, 1951).

EPILOGUE

252 thousands of Miocene-age fossils: Susan M. Kidwell, "Stratigraphic Condensation of Marine Transgressive Records: Origin of Major

Shell Deposits in the Miocene of Maryland," *Journal of Geology* 97, (1989): 1–24.

252 **Frederick True and Remington Kellogg:** Among other contributions, some examples include Frederick W. True, "Description of a New Genus and Species of Fossil Seal from the Miocene of Maryland," *Proceedings of the United States National Museum* 30 (1906): 835– 40; and Remington Kellogg, "Description of Two Squalodonts Recently Discovered in the Calvert Cliffs, Maryland; and Notes on the Shark-Toothed Cetaceans," *Proceedings of the United States National Museum* 62 (1923): 1–69.

253 **Kellogg worked on extensively:** Remington Kellogg published a massive eight-part monographic set of scientific papers featuring descriptions of fossil whales from the Calvert Formation at the end of his life. See, for example, Remington Kellogg, "A Hitherto Unrecognized Calvert Cetothere," *United States National Museum Bulletin* 247 (1968): 133–61.

254 **more than 23 million years ago:** R. Ewan Fordyce, "Cetacean Fossil Record," in *Encyclopedia of Marine Mammals*, 2nd ed. (2009), pp. 207–15.

254 **ten thousand squid meals:** Victor B. Scheffer, *The Year of the Whale* (New York: Scribner, 1969).

255 **sometimes do today:** Sperm whales around the world sleep this way, actually. It's not clear if they sleep with only half of their brain (called uni-hemispheric sleep) the way captive dolphins do—no one has conducted electroencephalography on a free-ranging, wild whale. Patrick J. O. Miller et al., "Stereotypical Resting Behavior of the Sperm Whale," *Current Biology* 18 (2008): R21–23.

SELECTED BIBLIOGRAPHY

Burnett, D. Graham. *The Sounding of the Whale: Science and Cetaceans in the Twentieth Century.* Chicago: University of Chicago Press, 2012.

Day, David. *Antarctica: A Biography.* Oxford, UK: Oxford University Press, 2013.

Estes, James A., Douglas P. DeMaster, Daniel F. Doak, Terrie M. Williams, and Robert L. Brownell Jr., eds. *Whales, Whaling, and Ocean Ecosystems.* Berkeley: University of California Press, 2007.

Gaskin, David E. *The Ecology of Whales and Dolphins.* New York: Heinemann, 1982.

George, Jean Craighead. *The Ice Whale.* New York: Dial Books for Young Readers, 2014.

Hoare, Philip. *The Whale: In Search of the Giants of the Sea.* London: Ecco, 2010.

Horwitz, Joshua. *War of the Whales: A True Story.* New York: Simon & Schuster, 2015.

Lilly, John C. *Man and Dolphin.* New York: Doubleday, 1961.

Lopez, Barry. *Arctic Dreams: Imagination and Desire in a Northern Landscape.* London: Picador, 1987.

Marx, Felix G., Olivier Lambert, and Mark D. Uhen. *Cetacean Paleobiology.* Hoboken, NJ: John Wiley and Sons, 2016.

Mowat, Farley. *A Whale for the Killing.* Toronto: McClelland & Stewart, 1972.

Norris, Kenneth S., ed. *Whales, Dolphins and Porpoises.* Berkeley: University of California Press, 1966.

Payne, Roger. *Among Whales.* New York: Scribner, 1995.

Philbrick, Nathaniel. *In the Heart of the Sea: The Tragedy of the Whaleship Essex.* New York: Viking, 2000.

Quammen, David. *The Reluctant Mr. Darwin: An Intimate Portrait of Charles Darwin and the Making of His Theory of Evolution.* New York: W. W. Norton, 2006.

Reiss, Diana. *The Dolphin in the Mirror: Exploring Dolphin Minds and Saving Dolphin Lives.* Wilmington, MA: Mariner Books, 2012.

Scammon, Charles M. *The Marine Mammals of the North-Western Coast of North America, Described and Illustrated: Together with an Account of the American Whale-Fishery.* San Francisco: John H. Carmany, 1874.

Scheffer, Victor B. *The Year of the Whale.* New York: Scribner, 1969.

Thewissen, J. G. M. *The Walking Whales: From Land to Water in Eight Million Years.* Berkeley: University of California Press, 2014.

Tønnessen, Johan Nicolay, and Arne Odd Johnsen. *The History of Modern Whaling.* Berkeley: University of California Press, 1982.

Whitehead, Hal. *Sperm Whales: Social Evolution in the Ocean.* Chicago: University of Chicago Press, 2003.

Zimmer, Carl. *At the Water's Edge: Macroevolution and the Transformation of Life.* New York: Free Press, 1998.

INDEX

Page numbers in *italics* indicate illustrations and charts.